BIRKHÄUSER

Advanced Courses in Mathematics
CRM Barcelona

Centre de Recerca Matemàtica

Silvia Bertoluzza
Silvia Falletta
Giovanni Russo
Chi-Wang Shu

Numerical Solutions of Partial Differential Equations

Birkhäuser
Basel · Boston · Berlin

Authors:

Silvia Bertoluzza
Istituto di Matematica Applicata e
Tecnologie Informatiche del C.N.R.
v. Ferrata 1
27100 Pavia
Italy
e-mail: silvia.bertoluzza@imati.cnr.it

Silvia Falletta
Dipartimento di Matematica
Politecnico di Torino
Corso Duca degli Abruzzi, 24
10129 Torino
Italy
e-mail: silvia.falletta@polito.it

Giovanni Russo
Dipartimento di Matematica ed Informatica
Università di Catania
Viale Andrea Doria 6
95125 Catania
Italy
e-mail: russo@dmi.unict.it

Chi-Wang Shu
Division of Applied Mathematics
Brown University
Providence, RI 02912
USA
e-mail: shu@dam.brown.edu

2000 Mathematical Subject Classification 35-99, 65-99

Library of Congress Control Number: 2008940758

Bibliographic information published by Die Deutsche Bibliothek
Die Deutsche Bibliothek lists this publication in the Deutsche Nationalbibliografie;
detailed bibliographic data is available in the Internet at <http://dnb.ddb.de>.

ISBN 978-3-7643-8939-0 Birkhäuser Verlag AG, Basel – Boston – Berlin

Part of Springer Science+Business Media
Printed on acid-free paper produced from chlorine-free pulp. TCF ∞
Printed in Germany

ISBN 978-3-7643-8939-0 e-ISBN 978-3-7643-8940-6

9 8 7 6 5 4 3 2 1 www.birkhauser.ch

Contents

II High Order Shock-Capturing Schemes for Balance Laws

Foreword

This book contains an expanded and smoothed version of lecture notes delivered by the authors at the Advanced School on *Numerical Solutions of Partial Differential Equations: New Trends and Applications*, which took place at the Centre de Recerca Matemàtica (CRM) in Bellaterra (Barcelona) from November 15th to 22nd, 2007.

The book has three parts. The first part, by Silvia Bertoluzza and Silvia Falletta, is devoted to the use of wavelets to derive some new approaches in the numerical solution of PDEs, showing in particular how the possibility of writing equivalent norms for the scale of Besov spaces allows to write down some new methods. The second part, by Giovanni Russo, provides an overview of the modern finite-volume and finite-difference shock-capturing schemes for systems of conservation and balance laws, with emphasis in giving a unified view of such schemes by identifying the essential aspects of their construction. In the last part Chi-Wang Shu gives a general introduction to the discontinuous Galerkin methods for solving some classes of PDEs, discussing cell entropy inequalities, nonlinear stability and error estimates.

The school that originated these notes was born with the objective of providing an opportunity for PhD students, recent PhD doctorates and researchers in general in fields of applied mathematics and engineering to catch up with important developments in the fields and/or to get in touch with state-of-the-art numerical techniques that are not covered in usual courses at graduate level.

We are indebted to the Centre de Recerca Matemàtica and its staff for hosting the advanced school and express our gratitude to José A. Carrillo (Institució Catalana de Recerca i Estudis Avançats – Universitat Autònoma de Barcelona), Rosa Donat (Universitat de Valéncia), Carlos Parés (Universidad de Málaga) and Yolanda Vidal (Universitat Politècnica de Catalunya) for the mathematical organisation of the course and for making it such a pleasant experience.

Part I

Wavelets and Partial Differential Equations

Silvia Bertoluzza and Silvia Falletta

Introduction

Wavelet bases were introduced in the late 1980s as a tool for signal and image processing. Among the applications considered at the beginning we recall applications in the analysis of seismic signals, the numerous applications in image processing – image compression, edge-detection, denoising, applications in statistics, as well as in physics. Their effectiveness in many of the mentioned fields is nowadays well established: as an example, wavelets are actually used by the US *Federal Bureau of Investigation* (or FBI) in their fingerprint database, and they are one of the ingredients of the new MPEG media compression standard. Quite soon it became clear that such bases allowed to represent objects (signals, images, turbulent fields) with singularities of complex structure with a low number of degrees of freedom, a property that is particularly promising when thinking of an application to the numerical solution of partial differential equations: many PDEs have in fact solutions which present singularities, and the ability to represent such a solution with as little as possible degrees of freedom is essential in order to be able to implement effective solvers for such problems. The first attempts to use such bases in this framework go back to the late 1980s and early 1990s, when the first simple adaptive wavelet methods [32] appeared. In those years the problems to be faced were basic ones. The computation of integrals of products of derivatives of wavelets – objects which are naturally encountered in the variational approach to the numerical solution of PDEs – was an open problem (solved later by Dahmen and Michelli in [25]). Moreover, wavelets were defined on $\mathbb{R}$ and on $\mathbb{R}^n$. Already solving a simple boundary value problem on $(0, 1)$ (the first construction of wavelets on the interval [20] was published in 1993) posed a challenge.

Many steps forward have been made since those pioneering works. In particular *thinking in terms of wavelets* gave birth to some new approaches in the numerical solution of PDEs. The aim of this course is to show some of these new ideas. In particular we want to show how one key property of wavelets (the possibility of writing equivalent norms for the scale of Besov spaces) allows to write down some new methods.

Chapter 1

What is a Wavelet?

Let us start by explaining what we mean by wavelets. There are in the literature many definitions of wavelets and wavelet bases, going from the more strict ones (a wavelet is the dilated and translated version of a *mother wavelet* satisfying a suitable set of properties) to more and more general definitions. The aim of this chapter is to review the classical definition of wavelets for $\mathbb{R}$ and then point out which of its properties can be retained when replacing $\mathbb{R}$ with a generic domain Ω.

1.1 Multiresolution Analysis

We start by introducing the general concept of multiresolution analysis in the univariate case.

Definition 1.1. A *Multiresolution Analysis* (MRA) of $L^2(\mathbb{R})$ is a sequence $\{V_j\}_{j\in\mathbb{Z}}$ of closed subspaces of $L^2(\mathbb{R})$ verifying:

i) the subspaces are nested: $V_j \subset V_{j+1}$ for all $j \in \mathbb{Z}$;

ii) the union of the spaces is dense in $L^2(\mathbb{R})$ and the intersection is null:

$$\overline{\bigcup_{j\in\mathbb{Z}} V_j} = L^2(\mathbb{R}), \qquad \bigcap_{j\in\mathbb{Z}} V_j = \{0\}; \tag{1.1}$$

iii) there exists a *scaling function* $\varphi \in V_0$ such that $\{\varphi(\cdot - k), k \in \mathbb{Z}\}$ is a Riesz's basis for V_0.

We recall that a set $\{e_k\}$ is a Riesz basis for its linear span in $L^2(\mathbb{R})$ if and only if the functions e_k are linearly independent and the following norm equivalence holds,

$$\Big\| \sum_k c_k e_k \Big\|^2_{L^2(\mathbb{R})} \simeq \sum_k |c_k|^2.$$

Here and in the following we use the notation $A \simeq B$ to signify that there exist positive constants c and C, independent of any relevant parameter, such that $cB \leq A \leq CB$. Analogously we will use the notation $A \lesssim B$ (resp. $A \gtrsim B$), meaning that $A \leq CB$ (resp. $A \geq cB$).

It is not difficult to check that the above properties imply that the set

$$\{\varphi_{j,k} = 2^{j/2}\varphi(2^j \cdot -k),\ k \in \mathbb{Z}\}$$

is a Riesz's basis for V_j, yielding a norm equivalence between the L^2-norm of a function in V_j and the ℓ^2-norm of the sequence of its coefficients with constants independent of j.

The inclusion $V_0 \subset V_1$ implies that the scaling function φ can be expanded in terms of the basis of V_1 through the following *refinement equation*

$$\varphi(x) = \sum_{k\in\mathbb{Z}} h_k \varphi(2x - k) \tag{1.2}$$

with $\{h_k\}_{k\in\mathbb{Z}} \in \ell^2(\mathbb{Z})$. The function φ is then said to be a *refinable function* and the coefficients h_k are called *refinement coefficients*.

Since $V_j \subset V_{j+1}$ it is not difficult to realize that an approximation f_{j+1} of a function f at level $j+1$ "contains" more information on f than the approximation f_j at level j. As an example, we can consider $f_j = P_j f$, where $P_j : L^2(\mathbb{R}) \to V_j$ denotes the $L^2(\mathbb{R})$-orthogonal projection onto V_j. Remark that $P_{j+1}P_j = P_j$ (a direct consequence of the nestedness of the spaces V_j). Moreover, we have that $P_j P_{j+1} = P_j$: f_{j+1} contains in this case all information needed to retrieve f_j. The idea is now to encode somehow the "loss of information" that we have when projecting f_{j+1} onto V_j. This is done by introducing the complement *wavelet space* W_j. In order to do that, we consider a more general framework, in which P_j is not necessarily the orthogonal projection and which yields the construction of a biorthogonal multiresolution analysis, as specified in the following section.

The Biorthogonal MRA

To be more general, let us start by choosing a sequence of uniformly bounded (not necessarily orthogonal) projectors $P_j : L^2(\mathbb{R}) \to V_j$ verifying the following properties:

$$P_j P_{j+1} = P_j, \tag{1.3}$$

$$P_j(f(\cdot - k2^{-j}))(x) = P_j f(x - k2^{-j}), \tag{1.4}$$

$$P_{j+1} f((2\cdot))(x) = P_j f(2x). \tag{1.5}$$

Remark again that the inclusion $V_j \subset V_{j+1}$ guarantees that $P_{j+1}P_j = P_j$. On the contrary, property (1.3) is not verified by general non-orthogonal projectors and expresses the fact that the approximation $P_j f$ can be derived from $P_{j+1} f$. Equations (1.4) and (1.5) require that the projector P_j respects the translation and dilation invariance properties (i) and (ii) of the MRA.

Since $\{\varphi_{0,k}\}$ is a Riesz's basis for V_0 we have that for $f \in L^2(\mathbb{R})$

$$P_0 f = \sum_k \alpha_k(f)\varphi_{0,k}$$

with $\alpha_k : L^2(\mathbb{R}) \to \mathbb{R}$ linear and continuous. By the Riesz's Representation Theorem, for each k, there exists an element $\tilde{\varphi}_{0,k} \in L^2(\mathbb{R})$ such that

$$\alpha_k(f) = \langle f, \tilde{\varphi}_{0,k} \rangle,$$

where we denote by $\langle \cdot, \cdot \rangle$ the L^2-scalar product. We have the following lemma:

Lemma 1.2. *The set $\{\tilde{\varphi}_{0,k},\ k \in \mathbb{Z}\}$ forms a Riesz's basis for the space $\tilde{V}_0 = P_0^*(L^2(\mathbb{R}))$ (where P_0^* denotes the adjoint of P_0). Moreover we have*

$$\tilde{\varphi}_{0,k}(x) = \tilde{\varphi}_{0,0}(x-k). \tag{1.6}$$

Proof. We start by remarking that since $\varphi_{0,n} \in V_0$, we have that

$$\varphi_{0,n} = P_0\varphi_{0,n} = \sum_k \langle \varphi_{0,n}, \tilde{\varphi}_{0,k} \rangle \varphi_{0,k},$$

and this implies

$$\langle \tilde{\varphi}_{0,n}, \varphi_{0,k} \rangle = \delta_{n,k}. \tag{1.7}$$

Remark that (1.7) implies that the functions $\tilde{\varphi}_{0,k}$ are linearly independent. By definition, since $\{\varphi_{0,k}\}$ is a Riesz's basis for V_0 there exist constants A and B such that

$$A\Big(\sum_k |\alpha_k|^2\Big)^{1/2} \leq \Big\| \sum_k \alpha_k \varphi_{0,k} \Big\|_{L^2(\mathbb{R})} \leq B\Big(\sum_k |\alpha_k|^2\Big)^{1/2}.$$

We have

$$\begin{aligned}
\Big\| \sum_k \xi_k \tilde{\varphi}_{0,k} \Big\|_{L^2(\mathbb{R})} &= \sup_{f \in L^2(\mathbb{R})} \frac{\langle f, \sum_k \xi_k \tilde{\varphi}_{0,k} \rangle}{\|f\|_{L^2(\mathbb{R})}} = \sup_{f \in L^2(\mathbb{R})} \frac{\sum_k \alpha_k(f)\xi_k}{\|f\|_{L^2(\mathbb{R})}} \\
&\lesssim \sup_{f \in L^2(\mathbb{R})} \frac{(\sum_k |\alpha_k(f)|^2)^{1/2}(\sum_k |\xi_k|^2)^{1/2}}{\|f\|_{L^2(\mathbb{R})}} \\
&\lesssim \sup_{f \in L^2(\mathbb{R})} \frac{\|P_0 f\|_{L^2(\mathbb{R})}(\sum_k |\xi_k|^2)^{1/2}}{\|f\|_{L^2(\mathbb{R})}} \lesssim \Big(\sum_k |\xi_k|^2\Big)^{1/2}.
\end{aligned} \tag{1.8}$$

Now we can write, using (1.7),

$$\sum_k |\xi_k|^2 = \Big\langle \sum_k \xi_k \varphi_{0,k}, \sum_k \xi_k \tilde{\varphi}_{0,k} \Big\rangle \tag{1.9}$$

$$\leq \Big\| \sum_k \xi_k \varphi_{0,k} \Big\|_{L^2(\mathbb{R})} \| \sum_k \xi_k \tilde{\varphi}_{0,k} \|_{L^2(\mathbb{R})}$$

$$\leq B \Big(\sum_k |\xi_k|^2 \Big)^{1/2} \Big\| \sum_k \xi_k \tilde{\varphi}_{0,k} \Big\|_{L^2(\mathbb{R})}. \tag{1.10}$$

The bound $(\sum_k |\xi_k|^2)^{1/2} \lesssim \| \sum_k \xi_k \tilde{\varphi}_{0,k} \|_{L^2(\mathbb{R})}$ follows by dividing both sides by $(\sum_k |\xi_k|^2)^{1/2}$. The set $\{\tilde{\varphi}_{0,k}\}$ is then indeed a Riesz basis for its linear span.

We now need to prove that $P_0^*(L^2(\mathbb{R})) = \text{span} < \tilde{\varphi}_{0,k} | k \in \mathbb{Z} >$. Let $f \in L^2(\mathbb{R})$. We have

$$\langle f, P_0^* \tilde{\varphi}_{0,n} \rangle = \langle \tilde{\varphi}_{0,n}, P_0 f \rangle = \Big\langle \tilde{\varphi}_{0,n}, \sum_k \langle f, \tilde{\varphi}_{0,k} \rangle \varphi_{0,k} \Big\rangle$$

$$= \sum_k \langle f, \tilde{\varphi}_{0,k} \rangle \langle \tilde{\varphi}_{0,n}, \varphi_{0,k} \rangle = \sum_k \langle f, \tilde{\varphi}_{0,k} \rangle \delta_{n,k} = \langle f, \tilde{\varphi}_{0,n} \rangle.$$

The arbitrariness of f implies that

$$P_0^* \tilde{\varphi}_{0,n} = \tilde{\varphi}_{0,n},$$

and thus $\text{span} < \tilde{\varphi}_{0,k} | k \in \mathbb{Z} > \subset P_0^*(L^2(\mathbb{R}))$. On the other hand, for all $f \in L^2(\mathbb{R})$ we have that $P_0^* P_0^* f = P_0^* f$ and therefore $f \in P_0^*(L^2(\mathbb{R}))$ implies that $f = P_0^* f$. Then, for $f \in P_0^*(L^2(\mathbb{R}))$ and $g \in L^2(\mathbb{R})$ we can write

$$\langle f, g \rangle = \langle (P_0^* f), g \rangle = \langle f, P_0 g \rangle = \Big\langle f, \sum_k \langle g, \tilde{\varphi}_{0,k} \rangle \varphi_{0,k} \Big\rangle = \Big\langle g, \sum_k \langle f, \varphi_{0,k} \rangle \tilde{\varphi}_{0,k} \Big\rangle.$$

Hence, thanks to the arbitrariness of g,

$$f = \sum_k \langle f, \varphi_{0,k} \rangle \tilde{\varphi}_{0,k},$$

which proves the reverse inclusion. In order to prove (1.6) we observe that

$$P_0(f(\cdot + n))(x) = \sum_k \langle f(\cdot + n), \tilde{\varphi}_{0,k} \rangle \varphi(x-k) = \sum_k \langle f(\cdot), \tilde{\varphi}_{0,k}(\cdot - n) \rangle \varphi(x-k),$$

$$P_0 f(x+n) = \sum_k \langle f(\cdot), \tilde{\varphi}_{0,k} \rangle \varphi(x-k+n).$$

Thanks to (1.4) we have that

$$\sum_k \langle f(\cdot), \tilde{\varphi}_{0,k}(\cdot - n) \rangle \varphi(x-k) = \sum_k \langle f(\cdot), \tilde{\varphi}_{0,k} \rangle \varphi(x-k+n),$$

and, since $\{\varphi_{0,k}\}$ is a Riesz's basis for V_0, implying that the coefficients (and in particular the coefficient of $\varphi_{0,0} = \varphi(x)$) are uniquely determined, this implies, for all $f \in L^2(\mathbb{R})$,

$$\langle f, \tilde{\varphi}_{0,0}(\cdot - n)\rangle = \langle f, \tilde{\varphi}_{0,n}\rangle,$$

that is, by the arbitrariness of f, $\tilde{\varphi}_{0,n} = \tilde{\varphi}_{0,0}(\cdot - n)$. □

Thanks to property (1.5) it is not difficult to prove the following lemma:

Lemma 1.3. *We have*

$$P_j f = \sum_k \langle \tilde{\varphi}_{j,k}, f\rangle \varphi_{j,k}, \qquad \text{with} \qquad \varphi_{j,k}(x) = 2^{j/2}\varphi(2^j x - k). \tag{1.11}$$

Moreover the set $\{\tilde{\varphi}_{j,k},\ k \in \mathbb{Z}\}$ forms a Riesz's basis for the subspace $\tilde{V}_j = P_j^*(L^2(\mathbb{R}))$ (where P_j^* denotes the adjoint of P_j).

Property (1.3) implies that the sequence $\tilde{V}_j$ is increasing, as stated by the following proposition:

Proposition 1.4. *The sequence $\tilde{V}_j$ satisfies $\tilde{V}_j \subset \tilde{V}_{j+1}$.*

Proof. Property (1.3) implies that $P_{j+1}^* P_j^* f = P_j^* f$. Now we have $f \in \tilde{V}_j$ implies $f = P_j^* f = P_{j+1}^* P_j^* f \in \tilde{V}_{j+1}$. □

Corollary 1.5. *The function $\tilde{\varphi} = \tilde{\varphi}_{0,0}$ is refinable.*

The above reasoning requires to choose a priori a sequence P_j, $j \in \mathbb{Z}$, of (oblique) projectors onto the subspaces V_j. A trivial choice is to define P_j as the $L^2(\mathbb{R})$-orthogonal projector. It is easy to see that all the required properties are satisfied by such a choice. In this case, since the $L^2(\mathbb{R})$-orthogonal projector is self adjoint, we have $\tilde{V}_j = V_j$, and the biorthogonal function $\tilde{\varphi}$ belongs itself to V_0. Clearly, in the case that $\{\varphi_{0,k}, k \in \mathbb{Z}\}$ is an orthonormal basis for V_0 (as in the Haar basis case of forthcoming Example I) we have that $\tilde{\varphi} = \varphi$. Another possibility would be to choose P_j to be the Lagrangian interpolation operator (as we will do for the Schauder basis in forthcoming Example IV). However this choice does not fit in our framework. In fact the interpolation is not well defined in $L^2(\mathbb{R})$. Moreover, depending on the characteristics of the spaces V_j, the existence of a uniquely defined Lagrangian interpolant of a given smooth function f is not automatically satisfied. Many other choices are possible in theory but quite difficult to construct in practice. The solution is then to go the other way round, and start by constructing the function $\tilde{\varphi}$. We then introduce the following definition:

Definition 1.6. A refinable function

$$\tilde{\varphi} = \sum_k \tilde{h}_k \tilde{\varphi}(2 \cdot - k) \in L^2(\mathbb{R}) \tag{1.12}$$

is dual to φ if

$$\langle \varphi(\cdot - k), \tilde{\varphi}(\cdot - l)\rangle = \delta_{k,l} \ k, l \in \mathbb{Z}.$$

Assuming that we have a refinable function $\tilde{\varphi}$ dual to φ, we can define the projector P_j as

$$P_j f = \sum_{k \in \mathbb{Z}} \langle f, \tilde{\varphi}_{j,k} \rangle \varphi_{j,k}.$$

P_j is a indeed projector: it is not difficult to check that $f \in V_j \Rightarrow P_j f = f$.

Remark 1.7. As it happened for the projector P_j, the dual refinable function $\tilde{\varphi}$ is not uniquely determined, once φ is given. Different projectors correspond to different dual functions. It is worth noting that P.G. Lemarié ([31]) proved that if φ is compactly supported, then there exists a dual function $\tilde{\varphi} \in L^2(\mathbb{R})$ which is itself compactly supported.

The dual of P_j

$$P_j^* f = \sum_{k \in \mathbb{Z}} \langle f, \varphi_{j,k} \rangle \tilde{\varphi}_{j,k}$$

is also an oblique projector onto the space $\mathrm{Im}(P_j^*) = \tilde{V}_j$, where

$$\tilde{V}_j = \mathrm{span} < \tilde{\varphi}_{j,k} | k \in \mathbb{Z} > .$$

It is not difficult to see that since $\tilde{\varphi}$ is refinable then the $\tilde{V}_j$'s are nested. This implies that for $j < l$ we have that $P_l^* P_j^* = P_j^*$ and $P_j P_l = P_j$.

Remark that there is a third approach that yields an equivalent structure. In fact, let the sequence V_j be given, and assume that we have a sequence $\tilde{V}_j$ of spaces such that the following inf-sup conditions hold uniformly in j:

$$\inf_{v_j \in V_j} \sup_{w_j \in \tilde{V}_j} \frac{\langle v_j, w_j \rangle}{\|v_j\|_{L^2(\mathbb{R})} \|w_j\|_{L^2(\mathbb{R})}} \gtrsim 1, \qquad \inf_{w_j \in \tilde{V}_j} \sup_{v_j \in V_j} \frac{\langle v_j, w_j \rangle}{\|v_j\|_{L^2(\mathbb{R})} \|w_j\|_{L^2(\mathbb{R})}} \gtrsim 1. \tag{1.13}$$

Then we can define a bounded projector $P_j : L^2(\mathbb{R}) \to V_j$ as $P_j v = v_j$, v_j being the unique element of V_j such that

$$\langle v_j, w_j \rangle = \langle v, w_j \rangle \qquad \forall w_j \in \tilde{V}_j.$$

It is not difficult to see that if the sequence $\tilde{V}_j$ is a multiresolution analysis (that is, if it satisfies the requirements of Definition 1.1), then the projector P_j satisfies properties (1.3), (1.4) and (1.5).

Remark 1.8. The uniform boundedness of the projector P_j and of its adjoint $\tilde{P}_j$ actually implies the validity of the two inf-sup conditions (1.13). We can see this by using the Fortin trick with P_j and $\tilde{P}_j$ as Fortin's projectors ([13]).

Wavelets

It is now straightforward to define a space W_j which complements V_j in V_{j+1} as

$$W_j = Q_j(L^2(\mathbb{R})), \qquad Q_j = P_{j+1} - P_j.$$

Remark that $Q_j^2 = Q_j$, that is Q_j is indeed a projector on W_j. W_j can also be defined as the kernel of P_j in V_{j+1}. We now need to construct a suitable basis for W_j. In order to do so, we introduce two sets of coefficients:

$$g_k = (-1)^k \tilde{h}_{1-k}, \qquad \tilde{g}_k = (-1)^k h_{1-k}, \qquad k \in \mathbb{Z}$$

and we introduce a pair of *dual wavelets*

$$\psi(x) = \sum_k g_k \varphi(2x - k), \qquad \tilde{\psi}(x) = \sum_k \tilde{g}_k \tilde{\varphi}(2x - k).$$

Clearly $\psi \in V_1$ and $\tilde{\psi} \in \tilde{V}_1$. The following theorem holds ([19]):

Theorem 1.9. *The wavelet functions ψ and $\tilde{\psi}$ satisfy*

$$\langle \psi, \tilde{\psi}(\cdot - k) \rangle = \delta_{0,k}$$

and

$$\langle \psi, \tilde{\varphi}(\cdot - k) \rangle = \langle \tilde{\psi}, \varphi(\cdot - k) \rangle = 0.$$

The projection operator Q_j can be expanded into

$$Q_j f = \sum_k \langle f, \tilde{\psi}_{j,k} \rangle \psi_{j,k} \, ,$$

and the functions $\psi_{j,k}$ constitute a Riesz's basis of W_j.

In summary we have a multiscale decomposition of V_j as

$$V_j = V_0 \oplus W_0 \oplus \cdots \oplus W_{j-1},$$

and for any function f, $P_j f$ in V_j can be expressed as

$$P_j f = \sum_k c_{j,k} \varphi_{j,k} = \sum_k c_{0,k} \varphi_{0,k} + \sum_{m=0}^{j-1} \sum_k d_{m,k} \psi_{m,k}, \tag{1.14}$$

with $d_{m,k} = \langle f, \tilde{\psi}_{m,k} \rangle$ and $c_{m,k} = \langle f, \tilde{\varphi}_{m,k} \rangle$. The approximation $P_j f$ is then decomposed as a coarse approximation at scale 0 plus a sequence of fluctuations at intermediate scales 2^{-m}, $m = 0, \ldots, j-1$. Both $\{\varphi_{j,k}, k \in \mathbb{Z}\}$ and $\{\varphi_{0,k}, k \in \mathbb{Z}\} \bigcup_{0 \le m < j} \{\psi_{m,k}, k \in \mathbb{Z}\}$ are bases for V_j and (1.14) expresses a change of basis. Thanks to the density property (1.1) for $j \to +\infty$, $P_j f$ converges to f in $L^2(\mathbb{R})$. Then, taking the limit for $j \to +\infty$ in (1.14) we obtain

$$f = \sum_k c_{0,k} \varphi_{0,k} + \sum_{m=0}^{+\infty} \sum_k d_{m,k} \psi_{m,k}.$$

We will see in the following that, under quite mild assumptions, the convergence is unconditional.

The Fast Wavelet Transform

The idea now is that, since all the information on $P_j f$ is encoded in the coefficients $c_{j,k}(f)$, we must be able to compute the coefficients $c_{j-1,k}(f)$ and $d_{j-1,k}(f)$ directly from the coefficients $c_{j,k}(f)$. Given $f_j = \sum_k c_{j,k}\varphi_{j,k} \in V_j$, we want to compute directly the coefficients of its approximation at the coarser scale

$$P_{j-1}f_j = \sum_k c_{j-1,k}\varphi_{j-1,k}.$$

We can do it thanks to the refinement equation, which gives us a "fine to coarse" discrete projection algorithm:

$$\begin{aligned} c_{j-1,k} &= 2^{(j-1)/2}\langle f_j, \tilde{\varphi}(2^{j-1}\cdot - k)\rangle = 2^{(j-1)/2}\Big\langle f_j, \sum_n \tilde{h}_n \tilde{\varphi}(2^j \cdot - 2k - n)\Big\rangle \\ &= \frac{1}{\sqrt{2}}\sum_n \tilde{h}_n c_{j,2k+n}. \end{aligned}$$

On the other hand, given the projection $P_{j-1}f_j = \sum_k c_{j-1,k}\varphi_{j-1,k}$ we are able to express it in terms of basis functions at the finer scale,

$$\begin{aligned} P_{j-1}f_j &= 2^{(j-1)/2}\sum_k c_{j-1,k}\varphi(2^{j-1}\cdot - k) \\ &= 2^{(j-1)/2}\sum_k c_{j-1,k}\sum_n h_n\varphi(2^j\cdot - 2k - n) \\ &= \frac{1}{\sqrt{2}}\sum_k\Big[\sum_n h_{k-2n}c_{j-1,n}\Big]\varphi_{j,k}. \end{aligned}$$

Analogously we have

$$Q_{j-1}f = \frac{1}{\sqrt{2}}\sum_k\Big[\sum_n g_{k-2n}d_{j-1,n}\Big]\varphi_{j,k}.$$

Since $P_j f = P_{j-1}f + Q_{j-1}f$ we immediately get

$$P_j f = \sum_k \frac{1}{\sqrt{2}}\Big[\sum_n h_{k-2n}c_{j-1,n} + \sum_n g_{k-2n}d_{j-1,n}\Big]\varphi_{j,k}.$$

Decomposition and reconstruction algorithm

In summary, the one level decomposition algorithm reads

$$c_{j,k} = \frac{1}{\sqrt{2}}\sum_k \tilde{h}_{k-2n}c_{j+1,k}, \qquad d_{j,k} = \frac{1}{\sqrt{2}}\sum_k \tilde{g}_{k-2n}c_{j+1,k}$$

while its inverse, the one level reconstruction algorithm can be written as

$$c_{j+1,k} = \frac{1}{\sqrt{2}}\Big[\sum_n h_{k-2n}c_{j,n} + \sum_n g_{k-2n}d_{j,n}\Big].$$

Once the one level decomposition algorithm is given, giving the coefficient vectors $(c_{j,k})_k$ and $(d_{j,k})_k$ in terms of the coefficient vector $(c_{j+1,k})_k$, we can iterate it to obtain $(c_{j-1,k})_k$ and $(d_{j-1,k})_k$ and so on until we get all the coefficients for the decomposition (1.14).

1.1.1 Example I: The Haar Basis

Let us consider the space V_j of piecewise constant functions with uniform mesh size $h = 2^{-j}$:

$$V_j = \{w \in L^2(\mathbb{R}) \text{ such that } w|_{I_{j,k}} \text{ is constant}\},$$

where we denote by $I_{j,k}$ the dyadic interval $I_{j,k} := (k2^{-j}, (k+1)2^{-j})$. An orthonormal basis for V_j is given by the family

$$\varphi_{j,k} := 2^{j/2}\varphi(2^j \cdot -k) \quad \text{with} \quad \varphi = \chi|_{(0,1)}.$$

Denoting by $P_j : L^2(\mathbb{R}) \to V_j$ the $L^2(\mathbb{R})$-orthogonal projection onto V_j, clearly we have

$$P_j f = \sum_k c_{j,k}(f)\varphi_{j,k}, \qquad c_{j,k}(f) = \langle f, \varphi_{j,k}\rangle.$$

The space W_j is the orthogonal complement of V_j in V_{j+1}:

$$V_{j+1} = W_j \oplus V_j, \qquad W_j \perp V_j,$$

and the $L^2(\mathbb{R})$-orthogonal projection $Q_j := P_{j+1} - P_j$ onto W_j verifies

$$\begin{aligned} Q_j f|_{I_{j+1,2k}} P_{j+1} f|_{I_{j+1,2k}} &- (P_{j+1}f|_{I_{j+1,2k}} + P_{j+1}f|_{I_{j+1,2k+1}})/2 \\ &= P_{j+1}f|_{I_{j+1,2k}}/2 - P_{j+1}f|_{I_{j+1,2k+1}}/2, \qquad (1.15) \\ Q_j f|_{I_{j+1,2k+1}} &= P_{j+1}f|_{I_{j+1,2k+1}} - (P_{j+1}f|_{I_{j+1,2k}} + P_{j+1}f|_{I_{j+1,2k+1}})/2 \\ &= -P_{j+1}f|_{I_{j+1,2k}}/2 + P_{j+1}f|_{I_{j+1,2k+1}}/2. \qquad (1.16) \end{aligned}$$

It is also not difficult to realize that we can expand $Q_j f$ as

$$Q_j f = \sum_k d_{j,k}(f)\psi_{j,k}, \qquad \text{with} \qquad \psi_{j,k} = 2^{j/2}\psi(2^j \cdot -k)$$

where

$$\psi := \chi_{(0,1/2)} - \chi_{(1/2,1)}.$$

Since the functions $\psi_{j,k}$ at fixed j are an orthonormal system, they do constitute an orthonormal basis for W_j. We have then

$$d_{j,k}(f) = \langle f, \psi_{j,k}\rangle.$$

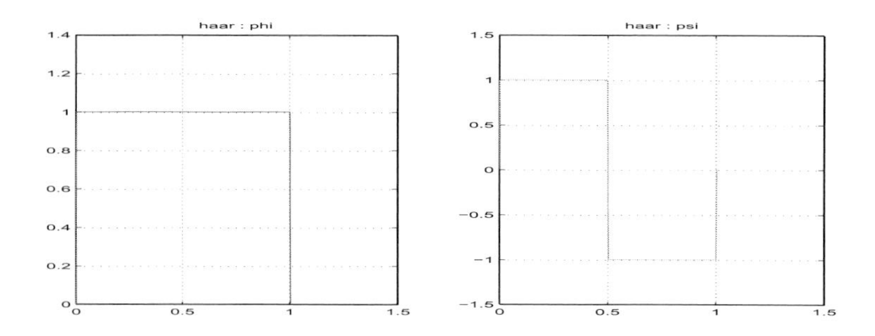

Figure 1.1: The scaling and wavelet functions φ and ψ generating the Haar basis.

1.1.2 Example II: B-Splines

We consider the space of splines of order $N+1$:

$$V_j = \{f \in L^2 \cap C^{N-1} \ : \ f|_{I_{j,k}} \in \mathbb{P}^N\}.$$

We construct a basis for V_j by defining the B-spline B_N of degree N recursively by

$$B_0 := \chi_{[0,1]},$$
$$B_N := B_0 * B_{N-1} = (*)^{N+1} \chi_{[0,1]},$$

where $*$ denotes the convolution product. The function B_N is supported in the interval $[0, N+1]$, it is refinable and the corresponding scaling coefficients are defined by

$$h_k = \begin{cases} 2^{-N} \begin{pmatrix} N+1 \\ k \end{pmatrix}, & 0 \le k \le N+1, \\ 0, \ \text{otherwise}. \end{cases}$$

The integer translates of the function $\varphi = B_N$ form a Riesz's basis for V_j. For N given it is possible to construct a whole class of compactly supported refinable functions dual to B_N. In particular for any given $\tilde{R}$ the dual function $\tilde{\varphi}$ can be chosen of regularity $\tilde{R}$. Figures 1.3 and 1.5 show the functions φ, $\tilde{\varphi}$, ψ and $\tilde{\psi}$ for $N = 1$, $\tilde{R} = 0$ and $N = 3$, $\tilde{R} = 1$ respectively.

1.1.3 Example III: Daubechies's Wavelets

Another important example of multiresolution analysis is given by Daubechies's orthonormal compactly supported wavelets. This is a class of MRA's such that the functions φ and ψ are both compactly supported and they generate by translations and dilations orthonormal bases for the spaces V_j and W_j. The projectors P_j are in this case L^2-orthogonal projectors. In this case the scaling function and the dual function coincide and we have $\tilde{V}_j = V_j$. The MRA $\{V_j\}$ can be chosen in such a way that φ and ψ have regularity R (with R arbitrary fixed number), the support of both φ and ψ increasing linearly with R. In the Daubechies wavelet construction the function φ is not explicitly given, but rather retrieved as the solution of the refinement equation (1.2) for which the coefficients h_k are given. It is beyond the scope of this course to give details about such a construction. We refer the interested reader to [28].

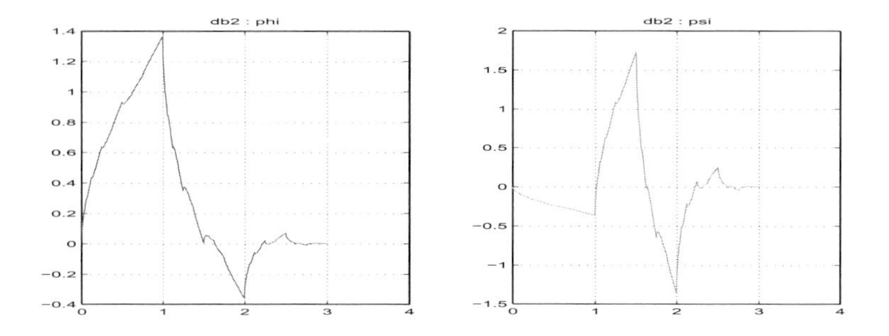

Figure 1.2: The scaling and wavelet functions φ and ψ generating a Daubechies' orthonormal wavelet basis.

1.1.4 Example IV: The Schauder Basis

It is interesting to consider for one minute an example that falls outside of the framework here described. Let us consider the space of continuous piecewise linear functions on a uniform mesh with meshsize 2^{-j},

$$V_j = \{w \in C^0(\mathbb{R}) : w \text{ is linear on } I_{j,k},\ k \in \mathbb{Z}\}.$$

We can easily construct a basis for V_j out of the dilated and translated of the "hat function":

$$V_j = \operatorname{span}\{\varphi_{j,k},\ k \in \mathbb{Z}\}, \qquad \text{with} \qquad \varphi_{j,k} := 2^{j/2}\varphi(2^j \cdot -k),$$

$$\varphi(x) = \max\{0, 1 - |x|\}.$$

This basis is a Riesz basis: $f \in V_j \cap L^2(\mathbb{R})$ implies

$$\Big\| \sum_k c_{j,k}\varphi_{j,k} \Big\|^2_{L^2(\mathbb{R})} \simeq \sum_k |c_{j,k}(f)|^2.$$

Remark that the hat function φ is the B-spline of order 1. The multiresolution analysis V_j itself falls then in the framework described in Section 1.1.2 and there exist a class of dual multiresolution analyses and of associated wavelets (in Figure

1.3 we see one of the possible dual functions). We want however to consider here a different, more straightforward, approach. We observe that $f_j \in V_j$ is uniquely determined by its point values at the mesh points $k2^{-j}$. Assuming that f is sufficiently regular we can consider consider the interpolant $f_j = P_j f$, with P_j denoting the Lagrange interpolation operator: $P_j : C^0(\mathbb{R}) \to V_j$ is defined by

$$P_j f(k2^{-j}) = f(k2^{-j}).$$

It is not difficult to realize that

$$f_j = P_j f = \sum_k c_{j,k}(f)\varphi_{j,k}, \qquad c_{j,k}(f) = 2^{-j/2} f(2^{-j}k).$$

Remark that P_j is a "projector" ($f \in V_j$ implies $P_j f = f$) but not an $L^2(\mathbb{R})$-bounded projector (it is not even well defined in L^2). Clearly we cannot find an L^2-function $\tilde{\varphi}$ allowing to write P_j in the form (1.11). However, if we allow ourselves to take $\tilde{\varphi}$ to be the Dirac's delta in the origin ($\tilde{\varphi} = \delta_{x=0}$) we see that the basic structure of the whole construction is preserved. Once again $V_j \subset V_{j+1}$ and $P_j f$ can be derived from $P_{j+1} f$ by interpolation,

$$2^{j/2} c_{j,k}(f) = f(k2^{-j}) = f(2k2^{-(j+1)}) = 2^{(j+1)/2} c_{j+1,2k}(f).$$

Also in this case we can compute the details that we loose when going from $P_{j+1} f$ to $P_j f$ as $Q_j f$, with $Q_j := P_{j+1} - P_j$.

Here the details $Q_j f$ at level j are not oscillating. Instead they vanish at the mesh points at level j. In fact $P_j f(k2^{-j}) = f(k2^{-j}) = P_{j+1} f(k2^{-j})$ implies

$$Q_j f(k2^{-j}) = 0.$$

We can then expand $Q_j f$ as

$$Q_j f = \sum d_{j,k}(f)\psi_{j,k}, \qquad \text{with} \qquad \psi_{j,k} = 2^{j/2}\psi(2\cdot -k),$$

where

$$\psi(x) = \varphi(2x-1).$$

This time, the "wavelets" $\psi_{j,k}$ are then simply those nodal functions at level 2^{j+2} associated to nodes that belong to the fine but not to the coarse grid.

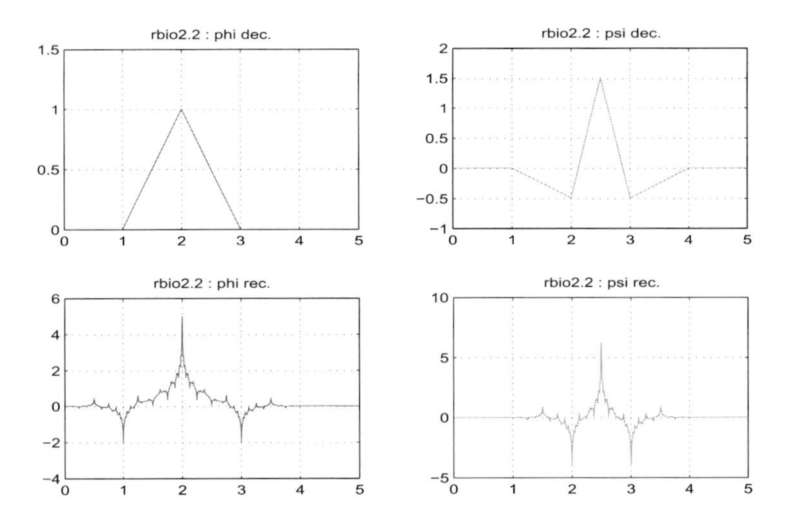

Figure 1.3: Scaling and wavelet functions φ and ψ for decomposition (top) and the duals $\tilde{\varphi}$ and $\tilde{\psi}$ for reconstruction (bottom) corresponding to the biorthogonal basis B2.2.

1.2 Beyond $L^2(\mathbb{R})$

What we just built for the space $L^2(\mathbb{R})$ is a complex structure consisting in:

a) two coupled multiresolution analyses V_j and $\tilde{V}_j$;

b) two sequences of adjont projectors $P_j : L^2(\mathbb{R}) \to V_j$ and $\tilde{P}_j = P_j^* : L^2(\mathbb{R}) \to \tilde{V}_j$, both verifying a commutativity property of the form (1.3);

c) two dual refinable functions φ and $\tilde{\varphi}$ (the scaling functions) which, by contraction and dilation generate bases for the V_j's and the $\tilde{V}_j$'s respectively, and that allow to write the two projectors P_j and P_j^* in the form (1.11);

d) a sequence of complement spaces W_j (and it is easy to build a second sequence $\tilde{W}_j$ of spaces complementing $\tilde{V}_j$ in $\tilde{V}_{j+1}$);

e) two functions ψ and $\tilde{\psi}$ which, by contraction and dilation generate bases for the W_j's and the $\tilde{W}_j$'s;

f) a fast change of basis algorithm, allowing to go back and forth from the coefficients of a given function in V_j with respect to the nodal basis $\{\varphi_{j,k},\ k \in \mathbb{Z}\}$ to the coefficients of the same function with respect to the hierarchical wavelet basis $\{\varphi_{0,k}, k \in \mathbb{Z}\} \bigcup_{m=0}^{j-1} \{\psi_{m,k}, k \in \mathbb{Z}\}$.

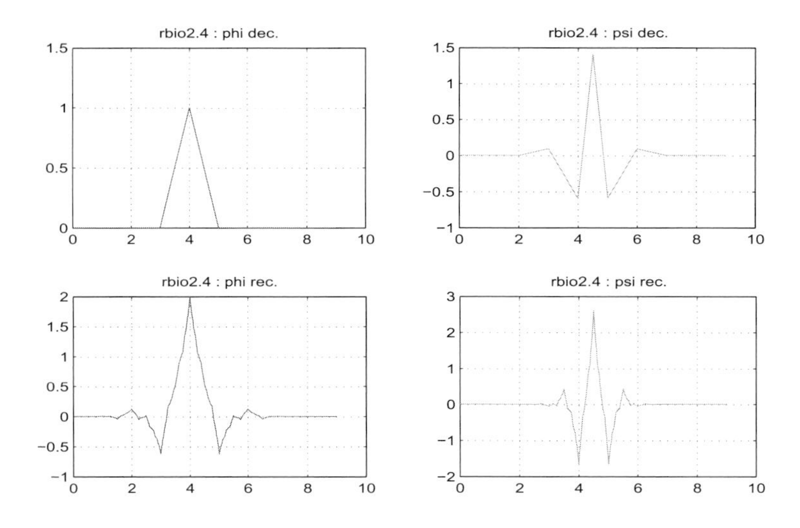

Figure 1.4: Example of a biorthogonal wavelet basis. Scaling and wavelet functions φ and ψ for decomposition (top) and the duals $\tilde{\varphi}$ and $\tilde{\psi}$ for reconstruction (bottom) corresponding to the basis B2.4. Remark that the scaling function for decomposition is the same as for the basis B2.2. In both cases V_j is the space of piecewise linears.

In view of the use of wavelets for the solution of PDEs, we would like to have a similar structure for more general domains, also in dimension greater than one. Actually, wavelets for $L^2(\mathbb{R}^n)$ are quite easily built by tensor product and we have basically the same structure as in dimension 1 (see, e.g., [16]). If, on the other hand, we want to build wavelets defined on general, possibly bounded, domains, it is clear that we have to somehow loosen the definition. In particular it is clear that for bounded domains we cannot ask for the translation and dilation invariance properties of the spaces V_j and the bases cannot possibly be constructed by contracting and translating a single function φ.

Let us then see which elements and properties of the above structure can be maintained when replacing the domain $\mathbb{R}$ with a general domain $\Omega \subseteq \mathbb{R}^n$. As we did for $\mathbb{R}$, we will start with a nested sequence $\{V_j\}_{j\geq 0}$, $V_j \subset V_{j+1}$, of closed subspaces of $L^2(\Omega)$, corresponding to discretizations with mesh-size 2^{-j}. We will still assume that the union of the V_j's is dense in $L^2(\Omega)$:

$$L^2(\Omega) = \overline{\bigcup_j V_j}. \tag{1.17}$$

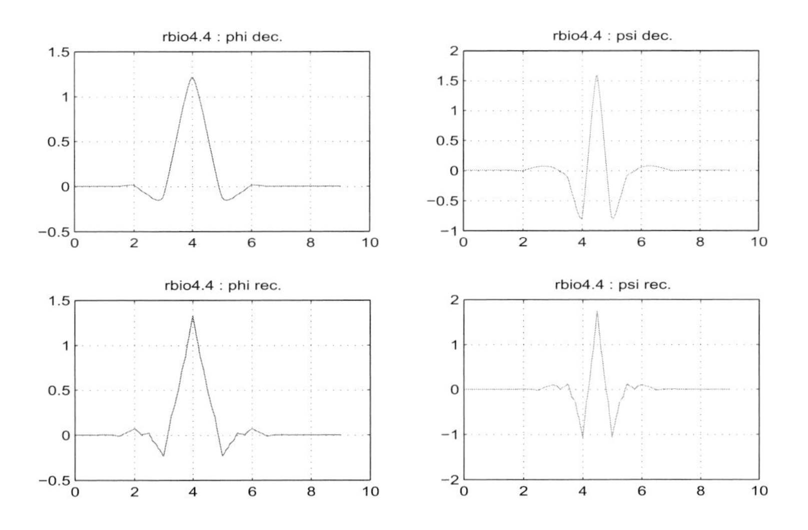

Figure 1.5: Example of a biorthogonal wavelet basis. Scaling and wavelet functions φ and ψ for decomposition (top) and the duals $\tilde{\varphi}$ and $\tilde{\psi}$ for reconstruction (bottom).

We will also assume that we have a Riesz's basis for V_j of the form $\{\varphi_\mu,\ \mu \in K_j\}$ such that

$$V_j = \operatorname{span}\{\varphi_\mu,\ \mu \in K_j\}, \qquad K_j \subseteq \{(j,k),\ k \in \mathbb{Z}^n\},$$

where K_j will denote a suitable set of multi-indexes (for $\Omega = \mathbb{R}$ the index set K_j will take the form $K_j = \{(j,k),\ k \in \mathbb{Z}\}$). Clearly, as already observed, it will not be possible to assume the existence of a single function φ such that all the basis functions φ_μ are obtained by dilating and translating φ. However remark that a great number of MRA's in bounded domains is built starting from an MRA for $L^2(\mathbb{R}^n)$ with scaling function φ compactly supported. In such a case, all the basis functions of the original MRA for $L^2(\mathbb{R}^n)$ whose support is strictly embedded in Ω are retained as basis functions for the V_j on Ω.

We now want to build a wavelet basis. To this aim we will need to introduce either a sequence of bounded projectors $P_j : L^2(\Omega) \to V_j$ satisfying $P_j P_{j+1} = P_j$ (note that $V_j = P_j(L^2(\Omega))$ and that $V_j \subset V_{j+1}$ implies $P_{j+1}P_j = P_j$) or, equivalently, a nested sequence of dual spaces $\tilde{V}_j$ satisfying the two inf-sup conditions mentioned in Section 1.1. Remark that, as it happens in the $L^2(\mathbb{R})$ case, choosing P_j is equivalent to choosing $\tilde{V}_j$. The existence of a biorthogonal Riesz's basis $\{\tilde{\varphi}_\mu,\ \mu \in K_j\}$ such that

$$\tilde{V}_j = P_j^*(L^2(\Omega)) = \operatorname{span}\{\tilde{\varphi}_\mu,\ \mu \in K_j\},$$

and such that

$$P_j f = \sum_{\mu \in K_j} \langle f, \tilde{\varphi}_\mu \rangle \varphi_\mu, \qquad P_j^* f = \sum_{\mu \in K_j} \langle f, \varphi_\mu \rangle \tilde{\varphi}_\mu$$

is easily deduced as in the $L^2(\mathbb{R})$ case (again, it will not generally be possible to obtain the basis functions $\tilde{\varphi}_\mu$ by dilations and translation of a single function $\tilde{\varphi}$).

As we did for $\mathbb{R}$ we can then introduce the difference spaces

$$W_j = Q_j(L^2(\Omega)), \qquad Q_j = P_{j+1} - P_j.$$

We will next have to construct a basis for W_j. This is in general a quite technical task, heavily depending on the particular characteristics of the spaces V_j and $\tilde{V}_j$. It's worth mentioning that, once again, if the MRA for Ω is built starting from an MRA for $L^2(\mathbb{R}^n)$ with compactly supported scaling function φ and if the wavelets themselves are compactly supported, then the basis for W_j will include all those wavelet functions on $\mathbb{R}$ whose support, as well as the support of the corresponding dual, are included in Ω. It is well beyond the scope of this book to go into the details of one or another construction of the basis for W_j. In any case, independently of the particular approach used, we will end up with a Riesz basis for W_j of the form $\{\psi_\lambda,\ \lambda \in \Lambda_j\}$ such that

$$W_j = \mathrm{span}\{\psi_\lambda,\ \lambda \in \Lambda_j\},$$

where Λ_j is again a suitable multi-index set with

$$\#(\Lambda_j) + \#(K_j) = \#(K_{j+1}).$$

At the same time we will end up with a Riesz basis for the dual spaces $\tilde{W}_j = (P_{j+1}^* - P_j^*)(L^2(\Omega))$:

$$\tilde{W}_j = \mathrm{span}\{\tilde{\psi}_\lambda,\ \lambda \in \Lambda_j\}.$$

The two bases can be chosen in such a way that they satisfy a biorthogonality relation

$$\langle \psi_\mu, \tilde{\psi}_{\mu'} \rangle = \delta_{\mu,\mu'}, \mu, \mu \in \Lambda_j,$$

so that the projection operator Q_j can be expanded as

$$Q_j f = \sum_{\lambda \in \Lambda_j} \langle f, \tilde{\psi}_\lambda \rangle \psi_\lambda.$$

Moreover it is not difficult to check that we have an orthogonality relation across scales:

$$\lambda \in \Lambda_j,\ \lambda' \in \Lambda_{j'},\ j \neq j' \Rightarrow \langle \psi_\mu, \tilde{\psi}_{\mu'} \rangle = 0, \qquad \mu \in K_{j'},\ j' \leq j \Rightarrow \langle \psi_\lambda, \tilde{\varphi}_\mu \rangle = 0.$$

In summary we have a multiscale decomposition of V_j as

$$V_j = V_0 \oplus W_0 \oplus \cdots \oplus W_{j-1},$$

and any function f_j in V_j can be expressed as

$$f_j = \sum_{\mu \in K_j} c_\mu \varphi_\mu = \sum_{\mu \in K_0} c_\mu \varphi_\mu + \sum_{m=0}^{j-1} \sum_{\lambda \in \Lambda_m} d_\lambda \psi_\lambda, \tag{1.18}$$

with $d_\lambda = \langle f, \tilde{\psi}_\lambda \rangle$ and $c_\mu = \langle f, \tilde{\varphi}_\mu \rangle$. For any $f \in L^2(\Omega)$ we can then write

$$P_j f = \sum_{\mu \in K_j} c_\mu \varphi_\mu = \sum_{\mu \in K_0} c_\mu \varphi_\mu + \sum_{m=0}^{j-1} \sum_{\lambda \in \Lambda_m} d_\lambda \psi_\lambda$$

with $d_\lambda = \langle f, \tilde{\psi}_\lambda \rangle$ and $c_\mu = \langle f, \tilde{\varphi}_\mu \rangle$. Since the density property (1.17) implies that

$$\lim_{j \to +\infty} \| f - P_j f \|_{L^2(\Omega)} = 0,$$

taking the limit as j goes to $+\infty$ and using the density of $\cup V_j$ in $L^2(\Omega)$ allows us to write

$$f = \sum_{\mu \in K_0} c_\mu \varphi_\mu + \sum_{j \geq 0} \sum_{\lambda \in \Lambda_j} \langle f, \tilde{\psi}_\lambda \rangle \psi_\lambda. \tag{1.19}$$

Remark 1.10. A general strategy to build bases with the required characteristics for $]0,1[^n$ out of the bases for $\mathbb{R}^n$ has been proposed in several papers [20],[1]. To actually build wavelet bases for general bounded domains, several strategies have been followed. Following the same strategy as for the construction of wavelet bases for cubes, *wavelet frames* (all the properties mentioned here hold but, for each j the elements set $\{\psi_\lambda, \lambda \in \Lambda_j\}$ is not linearly independent) for $L^2(\Omega)$ (Ω Lipschitz domain) can be constructed according to [17]. The most popular approach nowadays is domain decomposition: the domain Ω is split as the disjoint union of tensorial subdomains Ω_ℓ and a wavelet basis for Ω is constructed by suitably assembling wavelet bases for the Ω_ℓ's [14],[26],[21]. The construction is quite technical, since it is not trivial to retain in the assembling procedure the properties of the wavelets. Alternatively we can think of building wavelets for general domains directly, without starting from a construction on $\mathbb{R}$. This is for instance the case of finite element wavelets (see, e.g., [27]).

Chapter 2

The Fundamental Property of Wavelets

In the previous chapter we saw in some detail what a couple of biorthogonal multiresolution analyses is, and how this structure allows to build a wavelet basis. However we did not yet introduce the one property that makes of wavelets the powerful tool that they are and that probably is their fundamental characteristics: the simultaneous good localization in both space and frequency.

Simplifying the Notation

We put ourselves in the framework described in Section 1.2. Let us start by introducing a notation that will allow us to write the wavelet expansion in a more compact form. We start by setting

$$\Lambda_{-1} = K_0, \qquad \text{and for } \lambda \in \Lambda_{-1}, \quad \psi_\lambda = \varphi_\lambda.$$

The expansion (1.19) can be rewritten as

$$f = \sum_{j=-1}^{+\infty} \sum_{\lambda \in \Lambda_j} \langle f, \tilde{\psi}_\lambda \rangle \psi_\lambda.$$

We will see in the next section that, under quite mild assumptions on φ, $\tilde{\varphi}$, ψ and $\tilde{\psi}$, the convergence in the expansion (1.19) is unconditional. This will allow us to use an even more compact notation:

$$f = \sum_{\lambda \in \Lambda} \langle f, \tilde{\psi}_\lambda \rangle \psi_\lambda, \qquad \Lambda = \bigcup_{j=-1}^{+\infty} \Lambda_j. \tag{2.1}$$

Such formalism will also be valid for the case $\Omega = \mathbb{R}$, where we will have

$$\Lambda_j = K_j = \{(j,k), \quad k \in \mathbb{Z}\}.$$

For $\lambda = (j,k)$ we will have

$$\varphi_\lambda = \varphi_{j,k} = 2^{j/2}\varphi(2^j x - k), \qquad \tilde{\varphi}_\lambda = \tilde{\varphi}_{j,k} = 2^{j/2}\tilde{\varphi}(2^j x - k),$$

$$\psi_\lambda = \psi_{j,k} = 2^{j/2}\psi(2^j x - k), \qquad \tilde{\psi}_\lambda = \tilde{\psi}_{j,k} = 2^{j/2}\tilde{\psi}(2^j x - k).$$

Remark that, for $\lambda \in \Lambda_{-1}$, the functions $\psi_\lambda = \varphi_\lambda$ have a different behaviour from the actual wavelets, that is the functions ψ_λ, $\lambda \in \Lambda_j$ for $j \geq 0$. This observation leads us to introduce a second compact notation, which will be useful in the cases where we need to exploit such difference:

$$f = \sum_{\mu \in K_0} \langle f, \tilde{\varphi}_\mu \rangle \varphi_\mu + \sum_{\lambda \in \Lambda^0} \langle f, \tilde{\psi}_\lambda \rangle \psi_\lambda, \qquad \Lambda^0 = \bigcup_{j=0}^{+\infty} \Lambda_j.$$

2.1 The case $\Omega = \mathbb{R}$: The Frequency Domain Point of View vs. The Space Domain Point of View

As we saw in the previous chapter, in the classical construction of wavelet bases for $L^2(\mathbb{R})$ [33], all basis functions φ_λ, $\lambda \in K_j$ and ψ_λ, $\lambda \in \Lambda_j$ with $j \geq 0$, as well as their duals $\tilde{\varphi}_\lambda$ and $\tilde{\psi}_\lambda$, are constructed by translation and dilation of a single *scaling function* φ and a single *mother wavelet* ψ (resp. $\tilde{\varphi}$ and $\tilde{\psi}$). Clearly, the properties of the function ψ will transfer to the functions ψ_λ and will imply properties of the corresponding wavelet basis.

We will then make some assumptions on φ and ψ as well as on their duals $\tilde{\varphi}$ and $\tilde{\psi}$. The first assumption deals with *space localization*. In view of an application to the numerical solution of PDEs we make such an assumption in quite a strong form: we ask that there exists an $L > 0$ and an $\tilde{L} > 0$ such that

$$\operatorname{supp}\varphi \subseteq [-L, L] \quad \Longrightarrow \quad \operatorname{supp}\varphi_\lambda \subseteq [(k-L)/2^j, (k+L)/2^j], \tag{2.2}$$

$$\operatorname{supp}\tilde{\varphi} \subseteq [-\tilde{L}, \tilde{L}] \quad \Longrightarrow \quad \operatorname{supp}\tilde{\varphi}_\lambda \subseteq [(k-\tilde{L})/2^j, (k+\tilde{L})/2^j], \tag{2.3}$$

$$\operatorname{supp}\psi \subseteq [-L, L] \quad \Longrightarrow \quad \operatorname{supp}\psi_\lambda \subseteq [(k-L)/2^j, (k+L)/2^j], \tag{2.4}$$

$$\operatorname{supp}\tilde{\psi} \subseteq [-\tilde{L}, \tilde{L}] \quad \Longrightarrow \quad \operatorname{supp}\tilde{\psi}_\lambda \subseteq [(k-\tilde{L})/2^j, (k+\tilde{L})/2^j], \tag{2.5}$$

that is, both the wavelet ψ_λ ($\lambda = (j,k)$) and its dual $\tilde{\psi}_\lambda$ will be supported around the point $x_\lambda = k/2^j$, and the size of their support will be of the order of 2^{-j}.

Now let us consider the Fourier transform of ψ. Since ψ is compactly supported, by Heisenberg's indetermination principle, its Fourier transform $\hat{\psi}$ cannot be itself compactly supported. However we assume that it is localized in some weaker sense around the frequency 1. More precisely we assume that the following properties hold: there exist an integer $M > 0$ and an integer $R > 0$, with $M > R$, such that for $n = 0, \ldots, M$ and for s such that $0 \leq s \leq R$ one has

$$\text{a)} \quad \frac{d^n\hat{\psi}}{d\xi^n}(0) = 0, \qquad \text{and} \qquad \text{b)} \quad \int_{\mathbb{R}} (1 + |\xi|^2)^s |\hat{\psi}(\xi)|^2 \, d\xi \lesssim 1. \tag{2.6}$$

Analogously, for $\tilde{\psi}$ we assume that there exist an integer $\tilde{M} > 0$ and an integer $\tilde{R} > 0$ such that for $n = 0, \dots, \tilde{M}$ and for r such that $0 \leq s \leq \tilde{R}$ one has

$$\text{a)} \quad \frac{d^n \hat{\tilde{\psi}}}{d\xi^n}(0) = 0, \qquad \text{and} \qquad \text{b)} \quad \int_{\mathbb{R}} (1 + |\xi|^2)^s |\hat{\tilde{\psi}}(\xi)|^2 \, d\xi \lesssim 1. \tag{2.7}$$

The frequency localisation property (2.6) can be rephrased directly in terms of the function ψ, rather than in terms of its Fourier transform: in fact (2.6) is equivalent to

$$\int_{\mathbb{R}} x^n \psi(x) \, dx = 0, n = 0, \dots, M, \qquad \text{and} \qquad \|\psi\|_{H^s(\mathbb{R})} \lesssim 1, 0 \leq s \leq R, \tag{2.8}$$

which, by a simple scaling argument implies

$$\int_{\mathbb{R}} x^n \psi_\lambda(x) \, dx = 0, n = 0, \dots, M, \qquad \text{and} \qquad \|\psi_\lambda\|_{H^s(\mathbb{R})} \lesssim 2^{js}, 0 \leq s \leq R. \tag{2.9}$$

Analogously, we can write, for $\tilde{\psi}_\lambda$

$$\int_{\mathbb{R}} x^n \tilde{\psi}_\lambda(x) \, dx = 0, n = 0, \dots, \tilde{M}, \qquad \text{and} \qquad \|\tilde{\psi}_\lambda\|_{H^s(\mathbb{R})} \lesssim 2^{js}, 0 \leq s \leq \tilde{R}. \tag{2.10}$$

In the following we will also require the functions φ and $\tilde{\varphi}$ to have some frequency localization property or, equivalently, some smoothness. More precisely we will ask that for all s and $\tilde{s}$ such that, respectively, $0 \leq s \leq R$ and $0 \leq \tilde{s} \leq \tilde{R}$, we have that

$$\text{a)} \int_{\mathbb{R}} (1 + |\xi|^2)^s |\hat{\varphi}(\xi)|^2 \, d\xi \lesssim 1, \qquad \text{and} \qquad \text{b)} \quad \int_{\mathbb{R}} (1 + |\xi|^2)^{\tilde{s}} |\hat{\tilde{\varphi}}(\xi)|^2 \, d\xi \lesssim 1, \tag{2.11}$$

or, equivalently, that

$$\text{a)} \quad \varphi \in H^R(\mathbb{R}), \qquad \text{and} \qquad \text{b)} \quad \tilde{\varphi} \in H^{\tilde{R}}(\mathbb{R}). \tag{2.12}$$

Remark 2.1. Heisenberg's uncertainty principle states that a function cannot be arbitrarily well localized both in space and frequency. More precisely, introducing the *position uncertainty* Δx_λ and the *momentum uncertainty* $\Delta \xi_\lambda$ defined by

$$\Delta x_\lambda := \left(\int (x - x_\lambda)^2 |\psi_\lambda(x)|^2 \, dx \right)^{1/2},$$

$$\Delta \xi_\lambda := \left(\int (\xi - \xi_\lambda)^2 |\hat{\psi}_\lambda(\xi)|^2 \, d\xi \right)^{1/2}$$

with $x_\lambda = x_{j,k} = k/2^j$ and $\xi_\lambda = \xi_{j,k} \sim 2^j$ defined by $\xi_\lambda = \int_{\mathbb{R}} \xi |\hat{\psi}_\lambda(\xi)|^2 \, d\xi$, one necessarily has $\Delta x_\lambda \cdot \Delta \xi_\lambda \geq 1$. In our case $\Delta x_\lambda \cdot \Delta \xi_\lambda \lesssim 1$, that is wavelets are simultaneously localised in space and frequency nearly as well as possible.

The frequency localization property of wavelets (2.6) and (2.8) can be rephrased in yet a third way as a *local polynomial reproduction* property.

Lemma 2.2. *Let* (2.2), (2.3), (2.4) *and* (2.5) *hold. Then* (2.7a) *holds if and only if for all polynomials p of degree $d \le \tilde{M}$ we have*

$$p = \sum_k \langle p, \tilde{\varphi}_{j,k} \rangle \varphi_{j,k}. \tag{2.13}$$

Analogously (2.6a) *holds if and only if for all polynomials p of degree $d \le M$ we have*

$$p = \sum_k \langle p, \varphi_{j,k} \rangle \tilde{\varphi}_{j,k}. \tag{2.14}$$

Remark that the expressions on the right-hand side of both (2.13) and (2.14) are well defined pointwise thanks to assumptions (2.2), (2.3), (2.4) and (2.5).

Proof. Let us prove that (2.6a) implies (2.13) (the reverse being straightforward). Let p be a polynomial of degree lower or equal than $\tilde{M}$, and let $I =]a, b[\subset \mathbb{R}$ be any bounded interval. Let $\hat{I}_j =]a - (L + \tilde{L})/2^j, a - (L + \tilde{L})/2^j[$. Consider the $L^2(\mathbb{R})$-function $\tilde{p}$ coinciding with p in $\hat{I}_j$ and vanishing in $L^2(\mathbb{R}) \setminus \hat{I}_j$. It is not difficult to realize that, for $m \ge j$, if (m, k) is such that $\operatorname{supp} \psi_{m,k} \cap I \ne \emptyset$, then $\operatorname{supp} \tilde{\psi}_{m,k} \subseteq \bar{\hat{I}}_j$; analogously if $\operatorname{supp} \varphi_{j,k} \cap I \ne \emptyset$, then $\operatorname{supp} \tilde{\varphi}_{j,k} \subseteq \bar{\hat{I}}_j$.

The density property (1.1) implies that for all $\varepsilon > 0$ there exists an $m > j$ such that

$$\|\tilde{p} - P_m(\tilde{p})\|_{L^2(I)} \le \|\tilde{p} - P_m(\tilde{p})\|_{L^2(\mathbb{R})} \le \varepsilon.$$

We can now write, setting $r_m = \tilde{p} - P_m(\tilde{p})$,

$$p|_I = \tilde{p}|_I = P_m(\tilde{p})|_I + r_m|_I = P_j(\tilde{p}) + \sum_{\ell=j}^{m-1} Q_\ell(\tilde{p})|_I + r_m|_I.$$

Now, the definition of $\tilde{p}$ implies that if $\lambda \in \Lambda_\ell$ with $\ell \ge j$, then $\operatorname{supp} \psi_\lambda \cap I \ne \emptyset$. Then $\tilde{p} = p$ on $\operatorname{supp} \tilde{\psi}_\lambda$ and therefore by (2.10) we have that $\langle \tilde{p}, \tilde{\psi}_\lambda \rangle = 0$, which implies $Q_\ell \tilde{p}|_I = 0$ for all $\ell > j$. This yields

$$p|_I = P_j(\tilde{p})_I + r_m|_I.$$

The definition of $\tilde{p}$ also implies that, for $\mu \in K_j$, if $\operatorname{supp} \varphi_\mu \cap I \ne \emptyset$, then $\tilde{p} = p$ on $\operatorname{supp} \tilde{\varphi}_\mu$ and then

$$p|_I = P_j(\tilde{p})|_I + r_m|_I = \sum_k \langle p, \tilde{\varphi}_{j,k} \rangle \varphi_{j,k}|_I + r_m|_I,$$

from which we deduce

$$\|p - \sum_k \langle p, \tilde{\varphi}_{j,k} \rangle \varphi_{j,k}\|_{L^2(I)} \le \varepsilon.$$

Thanks to the arbitrariness of ε and of the interval I and since, due to the local support of $\tilde{\varphi}$ the sum on the right-hand side of (2.13) is locally a finite sum, we can write

$$p = \sum_k \langle p, \tilde{\varphi}_{j,k} \rangle \varphi_{j,k}.$$

Analogously, for all polynomials of degree less or equal than M we can write

$$p = \sum_k \langle p, \varphi_{j,k} \rangle \tilde{\varphi}_{j,k}. \qquad \square$$

Remark 2.3. By abuse of notation we write (2.13) and (2.14) in the form

$$P_j(p) = p, \qquad \tilde{P}_j(p) = p.$$

Before going on in seeing what the space-frequency localisation properties of the basis function ψ (and consequently of the wavelets ψ_λ's) imply, let us consider functions with a stronger frequency localisation. Let us then drop the assumption that ψ is compactly supported and assume instead that its Fourier transform verifies

$$\operatorname{supp}(\hat{\psi}) \subset [-2,-1] \cup [1,2], \qquad \text{and that} \qquad \operatorname{supp}(\hat{f}) \subset [-1,1] \quad \forall f \in V_0.$$

Since one can easily check that for $\lambda = (j,k)$, $j \geq 0$, $\operatorname{supp}(\hat{\psi}) \subset [-2^{j+1}, -2^j] \cup [2^j, 2^{j+1}]$, observing that on $\operatorname{supp}(\hat{\psi}_\lambda)$ we have $|\xi| \simeq 2^j$ and that, for $\lambda \in \Lambda_j$ and $\mu \in \Lambda_m$ with $m \neq j$, the measure of $\operatorname{supp}(\hat{\psi}_\lambda) \cap \operatorname{supp}(\hat{\psi}_\mu)$ is 0, one immediately obtains the following equivalence: letting $f = \sum_\lambda f_\lambda \psi_\lambda$,

$$\|f\|^2_{H^s(\mathbb{R})} = \int_{\mathbb{R}} (1+|\xi|^2)^s |\hat{f}(\xi)|^2 \, d\xi \simeq \sum_j 2^{2js} \Big\| \sum_{\lambda \in \Lambda_j} f_\lambda \hat{\psi}_\lambda \Big\|^2_{L^2(\mathbb{R})} \tag{2.15}$$

(remark that for $j = -1$ we have $2^{2js} = 2^{-2s} \simeq 1$, the constant in the equivalence depending on s). By taking the inverse Fourier transform on the right-hand side we immediately see that

$$\|f\|^2_{H^s(\mathbb{R})} \simeq \sum_j 2^{2js} \Big\| \sum_{\lambda \in \Lambda_j} f_\lambda \psi_\lambda \Big\|^2_{L^2(\mathbb{R})}.$$

If $\{\psi_\lambda,\ \lambda \in \Lambda_j\}$ is a Riesz basis for W_j, (2.15) implies then

$$\|f\|^2_{H^s(\mathbb{R})} \simeq \sum_{j \geq -1} 2^{2js} \sum_{\lambda \in \Lambda_j} |f_\lambda|^2. \tag{2.16}$$

If we only consider partial sums, we easily derive direct and inverse inequalities, namely:

$$\Big\| \sum_{j=-1}^{J} \sum_{\lambda \in \Lambda_j} f_\lambda \psi_\lambda \Big\|_{H^s(\mathbb{R})} \lesssim 2^{Js} \Big\| \sum_{j=-1}^{J} \sum_{\lambda \in \Lambda_j} f_\lambda \psi_\lambda \Big\|_{L^2(\mathbb{R})}, \tag{2.17}$$

and

$$\Big\| \sum_{j=J+1}^{\infty} \sum_{\lambda \in \Lambda_j} f_\lambda \psi_\lambda \Big\|_{L^2(\mathbb{R})} \lesssim 2^{-Js} \Big\| \sum_{j=J+1}^{\infty} \sum_{\lambda \in \Lambda_j} f_\lambda \psi_\lambda \Big\|_{H^s(\mathbb{R})}. \tag{2.18}$$

Properties (2.17) and (2.18) – which, as we saw, are easily proven if $\hat{\psi}$ is compactly supported, continue to hold, though their proof is less evident, in the case of ψ compactly supported, provided (2.6) and (2.8) hold. The same is true for property (2.16). More precisely we can prove the following inequalities:

Theorem 2.4. *For s with $0 \leq s \leq \tilde{M} + 1$, $f \in H^s(\mathbb{R})$ implies*

$$\|f - P_j f\|_{L^2(\mathbb{R})} \lesssim 2^{-sj} |f|_{H^s(\mathbb{R})}. \tag{2.19}$$

Analogously, for $0 \leq s \leq M + 1$, $f \in H^s(\mathbb{R})$ implies

$$\|f - \tilde{P}_j f\|_{L^2(\mathbb{R})} \lesssim 2^{-sj} |f|_{H^s(\mathbb{R})}. \tag{2.20}$$

Proof. By the polynomial reproduction property (2.13) we have that $P_j(p) = p$ (in the $L^2_{loc}(\mathbb{R})$ sense) for any polynomial p of degree less or equal than $\tilde{M}$. We split $\mathbb{R}$ as the union of dyadic intervals $\mathbb{R} = \bigcup I_{j,k}$. Let $p_{j,k}$ be polynomials to be chosen in the following. We have

$$\begin{aligned} \|f - P_j f\|_{L^2(I_{j,k})} &= \|f - p_{j,k} + P_j(p_{j,k}) - P_j f\|_{L^2(I_{j,k})} \\ &\leq \|f - p_{j,k}\|_{L^2(I_{j,k})} + \|P_j(f - p_{j,k})\|_{L^2(I_{j,k})}. \end{aligned}$$

Letting $\tilde{I}_{j,k} =](k - (L + \tilde{L}))2^{-j}, (k + 1 + (L + \tilde{L}))2^{-j}[$ it is not difficult to verify that, thanks to the space localization assumptions on φ and $\tilde{\varphi}$, we have the bound

$$\|P_j(f - p_{j,k})\|_{L^2(I_{j,k})} \leq \|f - p_{j,k}\|_{L^2(\tilde{I}_{j,k})}.$$

We can then choose $p_{j,k}$ to be the best degree $\tilde{M}$ polynomial approximation of f on $\tilde{I}_{j,k}$. This implies (see [15]) $\|f - p_{j,k}\|_{L^2(I_{j,k})} \leq 2^{-sj} |f|_{H^s(\tilde{I}_{j,k})}$. Squaring and summing up over all $k \in \mathbb{Z}$ we obtain (2.19). The bound (2.20) is proven by the same argument. □

Applying the above theorem to $g = f - P_j f$ and observing that $g - P_j g = g$ we immediately obtain the bound

$$\|(I - P_j) f\|_{L^2(\mathbb{R})} \lesssim 2^{-js} \|(I - P_j) f\|_{H^s(\mathbb{R})}. \tag{2.21}$$

Theorem 2.5 (Inverse Inequality). *For all $f \in V_j$ and for all r with $0 \leq r \leq R$ it holds that*

$$\|f\|_{H^r(\mathbb{R})} \lesssim 2^{jr} \|f\|_{L^2(\mathbb{R})}. \tag{2.22}$$

Analogously for all $f \in \tilde{V}_j$ and for all r with $0 \leq r \leq \tilde{R}$ we have

$$\|f\|_{H^r(\mathbb{R})} \lesssim 2^{jr} \|f\|_{L^2(\mathbb{R})}. \tag{2.23}$$

Proof. We prove the result for r integer, the general case follows by standard space interpolation techniques. We have $f = \sum c_{j,k}\varphi_{j,k}$. Now $\varphi_{j,k} = 2^{j/2}\varphi(2^j \cdot -k) \Rightarrow \varphi_{j,k}^{(m)} = 2^{jm}2^{j/2}\varphi^{(m)}(2^j \cdot -k)$. For $m \le r$ we can write

$$\begin{aligned}\|f^{(m)}\|_{L^2(\mathbb{R})} &= \Big\|\sum c_{j,k}\varphi_{j,k}^{(m)}\Big\|_{L^2(\mathbb{R})}\\ &= 2^{jm}\Big\|\sum c_{j,k}\varphi^{(m)}(\cdot - k)\Big\|_{L^2(\mathbb{R})}.\end{aligned}$$

We need then to show that

$$\Big\|\sum c_{j,k}\varphi^{(m)}(\cdot - k)\Big\|_{L^2(\mathbb{R})} \lesssim \|f\|_{L^2(\mathbb{R})}(\sim \|c_{j,\cdot}\|_{\ell^2}). \tag{2.24}$$

We use the compactness of $\operatorname{supp}\varphi^{(m)}$:

$$\Big\|\sum c_{j,k}\varphi^{(m)}(\cdot - k)\Big\|^2_{L^2(\mathbb{R})} = \sum\Big\|\sum c_{j,k}\varphi^{(m)}(\cdot - k)\Big\|^2_{L^2(I_{0,n})}. \tag{2.25}$$

Only a fixed finite number of terms contributes to the $L^2(I_{0,n})$-norm. We can then use the equivalence of all norms in finite dimension and obtain

$$\begin{aligned}\Big\|\sum c_{j,k}\varphi^{(m)}(\cdot - k)\Big\|^2_{L^2(I_{0,n})} &\lesssim \sup_{k\in\{n+1-L,\cdot,n+L\}} |c_{j,k}|^2\ \|\varphi^{(m)}\|^2_{L^2(I_{0,n})}\\ &\lesssim \sum_{k=n+1-L}^{n+L} |c_{j,k}|^2,\end{aligned} \tag{2.26}$$

which implies (2.24). □

All functions in W_j and $\tilde{W}_j$ verify both direct and inverse inequality

$$\begin{aligned} f\in W_j &\quad\Rightarrow\quad \|f\|_{H^r(\mathbb{R})} \simeq 2^{jr}\|f\|_{L^2(\mathbb{R})}, \quad r\in[0,R],\\ f\in \tilde{W}_j &\quad\Rightarrow\quad \|f\|_{H^r(\mathbb{R})} \simeq 2^{jr}\|f\|_{L^2(\mathbb{R})}, \quad r\in[0,\tilde{R}].\end{aligned}$$

By a duality argument it is not difficult to prove that similar inequalities hold for negative values of s. More precisely, for $f \in W_j$ and $s \in [0,\tilde{R}]$ we have, using the identity $\tilde{Q}_j = \tilde{P}_{j+1}(I - P_j)$ and the direct inequality (2.20),

$$\begin{aligned}\|f\|_{H^{-s}(\mathbb{R})} &= \sup_{g\in H^s(\mathbb{R})}\frac{\langle f,g\rangle}{\|g\|_{H^s(\mathbb{R})}} = \sup_{g\in H^s(\mathbb{R})}\frac{\langle f,\tilde{Q}_j g\rangle}{\|g\|_{H^s(\mathbb{R})}}\\ &\lesssim \sup_{g\in H^s(\mathbb{R})}\frac{\|f\|_{L^2(\mathbb{R})}\|(I-\tilde{P}_j)g\|_{L^2(\mathbb{R})}}{\|g\|_{H^s(\mathbb{R})}} \lesssim 2^{-js}\|f\|_{L^2(\mathbb{R})}.\end{aligned}$$

Conversely we can write

$$\|f\|_{L^2(\mathbf{R})} = \sup_{g\in L^2(\mathbb{R})}\frac{\langle f,\tilde{Q}_j g\rangle}{\|g\|_{L^2(\mathbb{R})}} \lesssim \frac{\|f\|_{H^{-s}(\mathbb{R})}\|\tilde{Q}_j g\|_{H^s(\mathbb{R})}}{\|g\|_{L^2(\mathbb{R})}} \lesssim 2^{js}\|f\|_{H^{-s}(\mathbb{R})}.$$

In summary we have

$$f \in W_j \Rightarrow \|f\|_{H^s(\mathbb{R})} \simeq 2^{js}\|f\|_{L^s(\mathbb{R})}, \qquad s \in [-\tilde{R}, R], \tag{2.27}$$

$$f \in \tilde{W}_j \Rightarrow \|f\|_{H^s(\mathbb{R})} \simeq 2^{js}\|f\|_{L^s(\mathbb{R})}, \qquad s \in [-R, \tilde{R}]. \tag{2.28}$$

Remark 2.6. Note that an inequality of the form (2.22) is satisfied by all functions whose Fourier transform is supported in the interval $[-2^J, 2^J]$, while an inequality of the form (2.21) is verified by all functions whose Fourier transform is supported in $(-\infty, -2^J] \cup [2^J, \infty)$. Such inequalities are inherently bound to the frequency localisation of the functions considered, or, to put it in a different way, to their more or less oscillatory behaviour. Saying that a function is "low frequency" means that such function does not oscillate too much. This translates in an inverse type inequality. On the other hand, saying that a function is "high frequency" means that it is purely oscillating, that is that it is locally orthogonal to polynomials (where the meaning of "locally" is related to the frequency) and this translates in a direct inequality. In many applications the two relations (2.21) and (2.22) can actually replace the information on the localisation of the Fourier transform. In particular this will be the case when we deal with functions defined on a bounded set Ω, for which the concept of Fourier transform does not make sense. Many of the things that can be proven for the case $\Omega = \mathbb{R}$ by using Fourier transform techniques, can be proven in an analogous way for bounded Ω by suitably using inequalities of the form (2.21) and (2.22).

A consequence of the validity of properties (2.27) and (2.28) is the possibility of characterizing, through the wavelet coefficients, the regularity of a function. We first observe that, since all the functions $\tilde{\psi}_\lambda$ have a certain regularity, namely $\tilde{\psi}_\lambda \in H^{\tilde{R}}(\mathbb{R})$, the Fourier development (2.1) makes sense (at least formally), provided f has enough regularity for $\langle f, \tilde{\psi}_\lambda \rangle$ to make sense, at least as a duality product, that is provided $f \in (H^{\tilde{R}}(\mathbb{R}))'$. In addition, using (2.27) and (2.28) it is not difficult to prove that the following inf-sup conditions hold uniformly in j for all $s \in [-\tilde{R}, R]$:

$$\inf_{w \in W_j} \sup_{\tilde{w} \in \tilde{W}_j} \frac{\langle w, \tilde{w} \rangle}{\|w\|_{H^s(\mathbb{R})} \|\tilde{w}\|_{H^{-s}(\mathbb{R})}} \gtrsim 1, \qquad \inf_{\tilde{w} \in \tilde{W}_j} \sup_{w \in W_j} \frac{\langle w, \tilde{w} \rangle}{\|w\|_{H^s(\mathbb{R})} \|\tilde{w}\|_{H^{-s}(\mathbb{R})}} \gtrsim 1;$$

in fact, as in Remark 1.8, two inf-sup conditions with respect to the $L^2(\mathbb{R})$-norm are deduced from the $L^2(\mathbb{R})$-boundedness of Q_j and $\tilde{Q}_j$ and then (2.27) and (2.28) are used. This implies that Q_j and $\tilde{Q}_j$ can be extended to operators acting on $H^{-\tilde{R}}$ and H^{-R}, respectively.

The properties of wavelets imply that given any function $f \in H^{-\tilde{R}}(\mathbb{R})$, by looking at behaviour of the $L^2(\mathbb{R})$-norm of $Q_j f$ as j goes to infinity and, more in detail, by looking at the absolute values of the wavelet coefficients $\langle f, \tilde{\psi}_\lambda \rangle$, it is possible to establish whether or not a function belongs to certain function spaces, and it is possible to write an equivalent norm for such function spaces in terms of the wavelet coefficients. More precisely we have the following theorem (see [33, 22]).

Theorem 2.7. *Let assumptions* (2.2), (2.3), (2.4), (2.5), (2.6), (2.7) *and* (2.11) *hold. Let* $f \in H^{-\tilde{R}}$ *and let* $s \in]-\tilde{R}, R[$. *Then* $f \in H^s(\mathbb{R})$ *if and only if*

$$\|f\|_s^2 = \sum_{\mu \in K_0} |\langle f, \tilde{\varphi}_\mu \rangle|^2 + \sum_{j \geq 0} \sum_{\lambda \in \Lambda_j} 2^{2js} |\langle f, \tilde{\psi}_\lambda \rangle|^2 < +\infty. \tag{2.29}$$

Moreover $\| \cdot \|_s$ *is an equivalent norm for* $H^s(\mathbb{R})$.

Proof. Thanks to the fact that the functions φ_μ, $\mu \in K_0$ and ψ_λ, $\lambda \in \Lambda_j$ constitute Riesz's bases for V_0 and W_j respectively, equation (2.29) is equivalent to

$$\|P_0 f\|_{L^2(\mathbb{R})}^2 + \sum_{j \geq 0} 2^{2js} \|Q_j f\|_{L^2(\mathbb{R})}^2 < +\infty. \tag{2.30}$$

We will at first show that if (2.30) holds, then $\sum_j Q_j f \in H^s(\mathbb{R})$. We start by observing that a scalar product for $H^s(\mathbb{R})$ is defined by $\langle (1-\Delta)^{s/2} \cdot, (1-\Delta)^{s/2} \cdot \rangle = \langle (1-\Delta)^{s/2+\varepsilon} \cdot, (1-\Delta)^{s/2-\varepsilon} \cdot \rangle$. This allows us to write

$$\Big\| \sum_j Q_j f \Big\|_{H^s(\mathbb{R})}^2 \leq 2 \sum_j \sum_{k>j} \|Q_j f\|_{H^{s+2\varepsilon}(\mathbb{R})} \|Q_k f\|_{H^{s-2\varepsilon}(\mathbb{R})} + \sum_j \|Q_j f\|_{H^s(\mathbb{R})}^2.$$

Thanks to the inverse inequalities we can then bound

$$\Big\| \sum_j Q_j f \Big\|_{H^s(\mathbb{R})}^2 \leq 2 \sum_j \sum_{k>j} 2^{2js} \|Q_j f\|_{L^2(\mathbb{R})} 2^{2ks} \|Q_k f\|_{L^2(\mathbb{R})} 2^{-2\varepsilon|j-k|} + \sum_j 2^{2js} \|Q_j f\|_{L^2(\mathbb{R})}^2.$$

The second sum is finite by assumption and the first sum can be bound by recalling that the convolution product is a bounded operator from $\ell^1 \times \ell^2$ to ℓ^2.

Let now $f \in H^s(\mathbb{R})$. We have, for N arbitrary,

$$\Big[\|P_0 f\|_{L^2(\mathbb{R})}^2 + \sum_{j=1}^N 2^{2js} \|Q_j f\|_{L^2(\mathbb{R})}^2 \Big]^2 = \langle f, \tilde{P}_0 P_0 f + \sum_j 2^{2sj} \tilde{Q}_j Q_j f \rangle^2 \tag{2.31}$$

$$\lesssim \|f\|_{H^s(\mathbb{R})}^2 \|\tilde{P}_0 P_0 f + \sum_{j=1}^N 2^{2sj} \tilde{Q}_j Q_j f\|_{H^{-s}(\mathbb{R})}^2.$$

Using the first part of the theorem we get

$$\Big\| \tilde{P}_0 P_0 f + \sum_{j=1}^N 2^{2sj} \tilde{Q}_j Q_j f \Big\|_{H^{-s}(\mathbb{R})}^2 \lesssim \|\tilde{P}_0 P_0 f\|_{L^2(\mathbb{R})}^2 + \sum_{j=1}^N 2^{-2sj} 2^{4sj} \|\tilde{Q}_j Q_j f\|_{L^2(\mathbb{R})}^2.$$

Dividing both sides of equation (2.31) by $\|P_0 f\|^2_{L^2(\mathbb{R})} + \sum_{j=1}^N 2^{2sj}\|Q_j f\|^2_{L^2(\mathbb{R})}$ we obtain

$$\|P_0 f\|^2_{L^2(\mathbb{R})} + \sum_{j=1}^{N} 2^{2sj}\|Q_j f\|^2_{L^2(\mathbb{R})} \lesssim \|f\|_{H^s(\mathbb{R})}.$$

The arbitrariness of N yields the thesis. □

For $s = 0$ we immediately obtain the following corollary.

Corollary 2.8. *If the assumptions of Theorem 2.29 hold, then $\{\psi_\lambda,\ \lambda \in \Lambda\}$ is a Riesz basis for $L^2(\mathbb{R})$.*

A more general result actually holds. In fact, letting $B^{s,p}_q(\mathbb{R}) := B^s_q(L^p(\mathbb{R}))$ denote the Besov space of smoothness order s with summability in L^p and third index q (see, e.g., [37]) we have the following theorem ([33, 22]).

Theorem 2.9. *Let the assumptions of Theorem 2.29 hold. Let $f \in H^{-\tilde{R}}$ and let $s \in]-\tilde{R}, R[$, $0 < p, q < +\infty$. Then $f \in B^{s,p}_q(\mathbb{R})$ if and only if*

$$\|f\|^q_{s,p,q} = \Big(\sum_{\mu\in K_0} |\langle f, \tilde{\varphi}_\mu\rangle|^p\Big)^{q/p} + \sum_j \Big(\sum_{\lambda\in\Lambda_j} 2^{pjs}2^{p(1/2-1/p)j}|\langle f, \tilde{\psi}_\lambda\rangle|^p\Big)^{q/p} < +\infty. \tag{2.32}$$

Moreover $\|\cdot\|_{s,p,q}$ is an equivalent norm for $B^{s,p}_q(\mathbb{R})$. An analogous result, in which the ℓ^p- (resp. ℓ^q-) norms are replaced by the ℓ^∞-norm, holds for either $p = +\infty$ or $q = +\infty$ or both.

2.2 The General Case: Ω Domain of $\mathbb{R}^d$

Let us now consider the general case of Ω being a (possibly bounded) Lipschitz domain of $\mathbb{R}^d$. The property of space localization can be easily stated also for wavelet bases on general domains.

Localisation in space. For each $\lambda \in \Lambda_j$ we have that

$$\operatorname{diam}(\operatorname{supp}\varphi_\lambda) \lesssim 2^{-j}, \qquad \text{and} \qquad \operatorname{diam}(\operatorname{supp}\tilde{\varphi}_\lambda) \lesssim 2^{-j}, \tag{2.33}$$

$$\operatorname{diam}(\operatorname{supp}\psi_\lambda) \lesssim 2^{-j}, \qquad \text{and} \qquad \operatorname{diam}(\operatorname{supp}\tilde{\psi}_\lambda) \lesssim 2^{-j}, \tag{2.34}$$

and for all $\mathbf{k} = (k_1, k_2, \dots, k_d) \in \mathbb{Z}^d$ there are at most K (resp. $\tilde{K}$) values of $\lambda \in \Lambda_j$ such that

$$\operatorname{supp}\psi_\lambda \cap \square_{j,\mathbf{k}} \neq \emptyset \qquad (\text{resp. } \operatorname{supp}\tilde{\psi}_\lambda \cap \square_{j,\mathbf{k}}\ ) \tag{2.35}$$

(where $\square_{j,\mathbf{k}}$ denotes the cube of centre $\mathbf{k}/2^j$ and side 2^{-j}). This last requirement is equivalent to asking that the basis functions at j fixed are uniformly distributed over the domain of definition. It avoids, for instance, that they accumulate somewhere.

Clearly, the concept of frequency in the classical sense and the definition of Fourier transform do not make sense in such framework. Still, we can ask that the basis functions have the same property as the basis functions for $L^2(\mathbb{R})$ in terms of oscillations. We will then assume that they satisfy an analogous relation to (2.9). More precisely, using $x \in \Omega$ and $\alpha \in \mathbb{N}^d$ the notation $x^\alpha = (x_1, \cdots, x_d)^{(\alpha_1, \cdots, \alpha_d)} = x_1^{\alpha_1} + \cdots + x_d^{\alpha_d}$, we assume that the basis functions ψ_λ verify, with $\alpha \in \mathbb{N}^d$,

$$\|\psi_\lambda\|_{H^s(\Omega)} \lesssim 2^{js}, 0 \le s \le R, \quad \text{and} \quad \sum_i \alpha_i \le M \Rightarrow \int_\Omega x^\alpha \psi_\lambda(x)\,dx = 0. \quad (2.36)$$

A similar relation is required to hold for the dual basis:

$$\|\tilde{\psi}_\lambda\|_{H^s(\Omega)} \lesssim 2^{js}, 0 \le s \le \tilde{R}, \quad \text{and} \quad \sum_i \alpha_i \le \tilde{M} \Rightarrow \int_\Omega x^\alpha \tilde{\psi}_\lambda(x)\,dx = 0. \quad (2.37)$$

Using an argument similar to the one that allowed us to obtain (2.13) and (2.14) we obtain that (2.36) and (2.37) together with the property of space localization yield that for all polynomials p of degree less or equal than $\tilde{M}$ (resp. less or equal that M) we have

$$p = \sum_{\mu \in K_j} \langle p, \tilde{\varphi}_\mu \rangle \varphi_\mu \qquad (\text{resp. } p = \sum_{\mu \in K_j} \langle p, \varphi_\mu \rangle \tilde{\varphi}_\mu). \quad (2.38)$$

Exactly as in the case $\Omega = \mathbb{R}$ this property allows us to prove that a direct type inequality holds.

Theorem 2.10 (Direct Inequality). *Assume that* (2.33), (2.34), (2.35), (2.36) *and* (2.37) *hold. For all s, $0 < s \le \tilde{M} + 1$, $f \in H^s(\Omega)$ implies*

$$\|f - P_j f\|_{L^2(\Omega)} \lesssim 2^{-js} \|f\|_{H^s(\Omega)}. \quad (2.39)$$

Analogously for all s, $0 < s \le M + 1$, $f \in H^s(\Omega)$ implies

$$\|f - \tilde{P}_j f\|_{L^2(\Omega)} \lesssim 2^{-js} \|f\|_{H^s(\Omega)}. \quad (2.40)$$

With a proof which is quite similar to the one of the analogous result on $\mathbb{R}$ it is also not difficult to prove that an inverse inequality holds. More precisely we have the following theorem.

Theorem 2.11 (Inverse Inequality). *Assume that* (2.33), (2.34), (2.35), (2.36) *and* (2.37) *hold. For r with $0 < r \le R$ it holds that for all $f \in V_j$*

$$\|f\|_{H^r(\Omega)} \lesssim 2^{jr} \|f\|_{L^2(\Omega)}. \quad (2.41)$$

Analogously, for r with $0 < r \le \tilde{R}$ it holds that for all $f \in \tilde{V}_j$

$$\|f\|_{H^r(\Omega)} \lesssim 2^{jr} \|f\|_{L^2(\Omega)}. \quad (2.42)$$

Analogously to what happens for $\mathbb{R}$ also here we can prove a norm equivalence for $H^s(\Omega)$ in terms of a suitable weighted ℓ^2-norm of the sequence of wavelet coefficients. More precisely the following theorem, which is proven similarly to Theorem 2.9, holds ([22]).

Theorem 2.12. *Assume that* (2.33), (2.34), (2.35), (2.36), (2.37), (2.41) *and* (2.42) *hold. Let* $f \in (H^{\tilde{R}}(\Omega))'$ *and let* $-\tilde{R} < s < R$. *Let*

$$\|f\|_s^2 = \sum_{\mu \in K_0} |\langle f, \tilde{\varphi}_\mu \rangle|^2 + \sum_j \sum_{\lambda \in \Lambda_j} 2^{2js} |\langle f, \tilde{\psi}_\lambda \rangle|^2. \tag{2.43}$$

Then, for $s \geq 0$, $\|f\|_s$ *is an equivalent norm for the space* $H^s(\Omega)$ *and* $f \in H^s(\Omega)$ *if and only if* $\|f\|_s$ *is finite; for negative* s, $\|f\|_s$ *is an equivalent norm for the space* $(H^{-s}(\Omega))'$ *and* $f \in (H^{-s}(\Omega))'$ *if and only if* $\|f\|_s$ *is finite.*

An analogous result holds for the dual multiresolution analysis, which allows us to characterize $H^s(\Omega)$, $s \geq 0$ and $(H^{-s}(\Omega))'$, $s < 0$, for $-R < s < \tilde{R}$.

Remark 2.13. All the wavelets mentioned until now (with the exception of the Schauder basis) satisfy the assumptions of Theorem 2.12 for suitable values of M, $\tilde{M}$, R, $\tilde{R}$.

Thanks to these norm equivalences we can then evaluate the Sobolev norms for spaces with negative and/or fractionary indexes by using simple operations, namely the evaluation of $L^2(\Omega)$-scalar products and the evaluation of an (infinite) sum. Moreover, it is easy to realize that the norm $\|\cdot\|_s$ is an hilbertian norm, induced by the scalar product

$$(f, g)_s = \sum_j 2^{2js} \sum_{\lambda \in \Lambda_j} \langle f, \tilde{\psi}_\lambda \rangle \langle f, \tilde{\psi}_\lambda \rangle. \tag{2.44}$$

As a consequence, equation (2.44) provides us with an equivalent scalar product for the Sobolev spaces $H^s(\Omega)$, $s \geq 0$ and, for $s < 0$, for the dual space $(H^{-s}(\Omega))'$, which, for negative or fractionary values of s, is more easily evaluated than the original one.

Also in this case a charachterization result for Besov spaces holds, as stated by the following theorem (see once again [22]).

Theorem 2.14. *Let all the assumptions of Theorem* 2.12 *hold. Let* $f \in H^{-\tilde{R}}$ *and let* $s \in]-\tilde{R}, R[$, $0 < p, q < +\infty$. *Then* $f \in B^{s,p}_q(\Omega)$ *if and only if*

$$\|f\|^q_{s,p,q} = \Big(\sum_{\mu \in K_0} |\langle f, \tilde{\varphi}_\mu \rangle|^p \Big)^{q/p} + \sum_j \Big(\sum_{\lambda \in \Lambda_j} 2^{pjs} 2^{p(d/2-d/p)j} |\langle f, \tilde{\psi}_\lambda \rangle|^p \Big)^{q/p} < +\infty. \tag{2.45}$$

Moreover the $\|\cdot\|_{s,p,q}$ *is an equivalent norm for* $B^{s,p}_q(\Omega)$. *An analogous result, in which the* ℓ^p- *(resp.* ℓ^q-*) norms are replaced by the* ℓ^∞-*norm, holds for either* $p = +\infty$ *or* $q = +\infty$ *or both.*

A norm equivalence resulting from the above theorem which will be of interest is the case in which the expression on the left-hand side of (2.45) is the ℓ^τ-norm of the sequence of wavelet coefficients. This happens provided s, p and q are related by the relation

$$p = q = \tau, \qquad s = d/\tau - d/2. \tag{2.46}$$

If we consider the space $B^{s,\tau}_\tau$ we have the equivalent norm

$$\|f\|_{B^{s,\tau}_\tau(\Omega)} \simeq \|f\|_{s,\tau,\tau} = \Big(\sum_{\lambda\in\Lambda} |\langle f, \tilde{\psi}_\lambda\rangle|^\tau\Big)^{1/\tau}. \tag{2.47}$$

In the following we will see that the Besov spaces $B^{s,\tau}_\tau$ play a key role in the analysis on *nonlinear approximation* in $L^2(\Omega)$.

2.3 The Issue of Boundary Conditions

When aiming at using wavelet bases for the numerical solution of PDEs, one has to take into account the issue of boundary conditions. If, for instance, in the equation considered, essential boundary conditions (for example $u = g$) need to be imposed on a portion Γ_e of the boundary $\Gamma = \partial\Omega$ of the domain Ω of definition of the problem, we want that the basis functions ψ_λ, $\lambda \in \Lambda$, satisfy themselves the corresponding homogeneous boundary conditions on Γ_e (that is, in the example mentioned above, $\psi_\lambda = 0$ on Γ_e). Depending on the projectors P_j, the dual wavelets $\tilde{\psi}_\lambda$ however will not need to satisfy the same homogeneous boundary conditions, though this might be the case (if for instance the projector P_j is chosen to be the $L^2(\Omega)$-orthogonal projector). Depending on whether the ψ_λ and the $\tilde{\psi}_\lambda$ satisfy or not some homogeneous boundary conditions, the same boundary conditions will be incorporated in the spaces that we will be able to characterize through such functions. It is not the aim of this book to go into details but only to give an idea on the kind of results that hold. To fix the ideas let us then consider the case of $\Gamma_e = \Gamma$ and of Dirichlet boundary condition of order zero, namely $u = 0$ on Γ and let us concentrate on the characterization of Sobolev spaces. If, for all $\lambda \in \Lambda$, $\psi_\lambda = 0$ on Γ, then (2.43) will hold provided f belongs to the $H^s(\Omega)$ closure of $H^s(\Omega) \cap H^1_0(\Omega)$, that we will denote $\mathcal{H}^s_0(\Omega)$. If all the $\tilde{\psi}_\lambda$'s satisfy $\tilde{\psi}_\lambda = 0$, we cannot hope to characterize (through scalar products with such functions) the space $(H^s(\Omega))'$, but only the space $(\mathcal{H}^s_0(\Omega))'$. In particular, provided for all $\lambda \in \Lambda$ $\tilde{\psi}_\lambda = 0$ on Γ, for $s = -1$ we will have a characterization of the form

$$\|f\|^2_{H^{-1}(\Omega)} \simeq \Big(\sum_j 2^{-2j} \sum_{\lambda\in\Lambda_j} |\langle f, \tilde{\psi}_\lambda\rangle|^2\Big)^{1/2}. \tag{2.48}$$

Then, again, an equivalent $H^{-1}(\Omega)$ scalar product can be defined as

$$(f, g)_{-1} = \sum_j 2^{-2j} \sum_{\lambda\in\Lambda_j} \langle f, \tilde{\psi}_\lambda\rangle\langle f, \tilde{\psi}_\lambda\rangle. \tag{2.49}$$

Clearly, if for all $\lambda \in \Lambda$ the ψ_λ's satisfy an homogeneous boundary condition, we can expect a direct inequality of the form (2.39) to hold only if we assume that the function f to approximate satisfies itself the same homogeneous boundary conditions.

Chapter 3

Wavelets for Partial Differential Equations

The property of wavelet characterization of Sobolev and Besov spaces that we saw in the previous section is quite a powerful tool. In the next sections we will see how we can take advantage of such a property in the design of new efficient methods for the solution of PDEs. Let us assume from now on that we have a couple of multiresolution analyses V_j and $\tilde{V}_j$ satisfying all space and frequency localization assumptions of Section 2.2.

3.1 Wavelet Preconditioning

The first example of application of wavelets to the numerical solution of PDEs is the construction of optimal preconditioners for elliptic partial differential and pseudo differential equations [30, 23].

We consider the following framework: let V be an Hilbert space and let $A : V \to V'$ be a (pseudo) differential operator. Let $a : V \times V \to \mathbb{R}$ be the corresponding bilinear form:

$$a(u, v) := \langle Au, v \rangle.$$

We consider the following abstract problem.

Problem 3.1. Given $f \in V'$, find $u \in V$ solution to

$$Au = f.$$

We consider here the variational formulation: find $u \in V$ s.t.

$$a(u, v) = \langle f, v \rangle, \qquad \forall v \in V. \tag{3.1}$$

We make the following assumptions:

- a is *continuous*: for all $u, v \in V$

$$|a(u,v)| \lesssim \|u\|_V \|v\|_V. \tag{3.2}$$

- a is *coercive*: for all $v \in V$

$$a(v,v) \gtrsim \|v\|_V^2. \tag{3.3}$$

Many examples exist of problems that fall in such a framework: we can mention the Laplace equation with Dirichlet B.C.: the equations describing linear elasticity in different frameworks, the Helmholtz equation, different types of integral equations (single-layer potential, double-layer potential), and many others.

Let now $U_h \subset V$ be a closed subspace of V. The Galerkin method for solving Problem 3.1 is defined as follows: find $u_h \in U_h$ s.t.

$$a(u_h, v_h) = \langle f, v_h \rangle, \qquad \forall\, v_h \in U_h. \tag{3.4}$$

It is well known that an error estimate can be written down in terms of the best approximation error for the continuous solution u. Under the above assumptions Cea's lemma states in fact that

$$\|u - u_h\|_V \lesssim \inf_{v_h \in V_h} \|u - v_h\|_V.$$

If $V = H^s(\Omega)$ with $s \in]-\tilde{R}, R[$, we can take $U_h = V_j$ for some $j > 0$. Combining Cea's Lemma with the direct estimate (2.39) we immediately obtain that, if the continuous solution u to problem (3.1) verifies $u \in H^{s+t}(\Omega)$ with $s + t \leq \tilde{M} + 1$, then the error verifies

$$\|u - u_h\|_{H^s(\Omega)} \lesssim 2^{-jt} \|u\|_{H^{s+t}(\Omega)}.$$

In order to practically compute the approximate solution u_h we can proceed as usual: choose a basis for the approximation space, express the approximate solution u_h and the test function v_h as a linear combination of the basis functions, and reduce the equation (3.4) to a linear system. In our case we have two available bases for $U_h = V_j$, so we get two equivalent linear systems

$$\text{a) } Ac = f, \qquad \text{and} \qquad \text{b) } Rd = g, \tag{3.5}$$

the approximate solution u_h being

$$u_h = \sum_{\mu \in K_j} c_\mu \varphi_\mu = \sum_{\lambda \in \Lambda_h} d_\lambda \psi_\lambda, \quad \Lambda_h = \bigcup_{m=-1}^{j-1} \Lambda_m.$$

In particular we will have two different stiffness matrices

$$S = (s_{\mu,\mu'}), \quad s_{\mu,\mu'} = a(\varphi_{\mu'}, \varphi_\mu), \qquad \text{and} \qquad R = (r_{\lambda,\lambda'}), \quad r_{\lambda,\lambda'} = a(\psi_{\lambda'}, \psi_\lambda).$$

A problem that is encountered in the numerical solution of (pseudo) differential equations is that the condition number of the stiffness matrices A and R is usually quite high. Recall that the number of iterations needed for solving numerically a linear system grows with such a condition number. Let us recall the definition of condition number.

Definition 3.2. Let A be an invertible matrix. The *condition number* of A is defined as

$$\kappa(A) = \|A\|_2 \cdot \|A^{-1}\|_2.$$

A remedy to ill conditioning of A is given by *preconditioning* [35]. Given A with $\mathrm{cond}(A) \gg 1$ the idea is to find P such that

$$\kappa(PA) \simeq 1; \tag{3.6}$$

then solve

$$PAx = Pf. \tag{3.7}$$

Remark that if the matrix A is symmetric one would like to be able to exploit this in the choice of the solution method for the linear system (3.7). On the other hand, the matrix PA is generally non-symmetric (even when both A and P have this property). The remedy is to resort to the so called *split preconditioning*: given A with $\mathrm{cond}(A) \gg 1$, find E such that

$$\kappa(E^T AE) \simeq 1. \tag{3.8}$$

We can then apply our favorite solver to the (symmetric, if A is symmetric) linear system

$$E^T AEy = E^T f, \tag{3.9}$$

and then retrieve x as $x = Ey$.

Remark 3.3. If P is a symmetric positive definite matrix such that (3.6) holds, and if E is such that

$$E^T E = P, \tag{3.10}$$

then E satisfies (3.8). Though in this case a matrix E satisfying (3.10) always exists, its actual computation might be quite expensive. In such a case it is possible to implement the conjugate gradient method for the linear system (3.9) in such a way that only the multiplication by P is needed ([36]). As we will see this is not the case of the wavelet preconditioner, where E is directly available.

We start by considering the stiffness matrix R with respect to the wavelet basis and we consider a symmetric diagonal preconditioner. This is equivalent to rescaling the basis functions ψ_λ. More precisely let for $\lambda \in \Lambda_j$, $\omega_\lambda = 2^{js}$ and set $\check{\psi}_\lambda = \omega_\lambda^{-1}\psi_\lambda$. The coefficients with respect to the new basis are $\check{u}_\lambda = \omega_\lambda u_\lambda$:

$$u = \sum_{\lambda \in \Lambda_h} u_\lambda \psi_\lambda = \sum_{\lambda \in \Lambda_h} \omega_\lambda u_\lambda \omega_\lambda^{-1} \psi_\lambda = \sum_{\lambda \in \Lambda_h} \check{u}_\lambda \check{\psi}_\lambda.$$

Letting D be a diagonal matrix with diagonal entries $D_{\lambda,\lambda} = \omega_\lambda$, the stiffness matrix w.r.t. the new basis becomes then

$$\check{R} = D^{-1} R D^{-1}.$$

We have the following theorem ([30, 23]):

Theorem 3.4. *Uniformly in j*

$$\kappa(\check{R}) = \kappa(D^{-1} R D^{-1}) \lesssim 1.$$

Proof. We start by observing that the norm equivalences of the previous section can be rewritten as

$$\Big\| \sum_\lambda f_\lambda \psi_\lambda \Big\|_{H^s(\Omega)} \simeq \|(\omega_\lambda f_\lambda)_\lambda\|_{\ell^2}, \qquad \Big\| \sum_\lambda g_\lambda \tilde{\psi}_\lambda \Big\|_{H^{-s}(\Omega)} \simeq \|(\omega_\lambda^{-1} g_\lambda)_\lambda\|_{\ell 2},$$

(Remark that $g_\lambda = \langle g, \psi_\lambda \rangle$). We observe that:

1. since $g \in H^{-s}(\Omega)$, the right-hand side $\underline{g} = (g_\lambda)_\lambda = (\langle g, \psi_\lambda \rangle)_\lambda$ verifies

$$\|D^{-1} \underline{g}\|_{\ell^2} \simeq \|g\|_{H^{-s}(\Omega)};$$

2. since $u \in H^s(\Omega)$, the unknown coefficient vector $\underline{u} = (u_\lambda)_\lambda = (\langle u, \tilde{\psi}_\lambda \rangle)_\lambda$ satisfies

$$\|D \underline{u}\|_{\ell^2} \simeq \|u\|_{H^s(\Omega)}.$$

In order to prove the theorem we need to bound $\|\check{R}\|$ and $\|\check{R}^{-1}\|$. Now we have (with $v = \sum_\lambda x_\lambda \check{\psi}_\lambda$, $w = \sum_\lambda y_\lambda \check{\psi}_\lambda$)

$$\|\check{R}\| = \sup_x \sup_y \frac{y^T \check{R} x}{\|x\|_{\ell^2} \|y\|_{\ell^2}} = \sup_{v \in V_j} \sup_{w \in V_j} \frac{a(v,w)}{\|v\|_{H^s(\Omega)} \|w\|_{H^s(\Omega)}} \lesssim 1. \tag{3.11}$$

We now observe that setting $y = \check{R}^{-1} x$, the function $u = \sum_\lambda y_\lambda \check{\psi}_\lambda$ is the Galerkin projection onto V_j of the solution u to the equation

$$a(u,v) = \langle v, f \rangle, \quad \forall v \in V_j$$

with $f = \sum_\lambda x_\lambda \check{\psi}_\lambda$, and, thanks to the properties of a we have the bound

$$\|y\|_{\ell^2} \lesssim \|u\|_{H^s(\Omega)} \lesssim \|f\|_{H^{-s}(\Omega)} \lesssim \|x\|_{\ell^2}.$$

This implies $\|\check{R}^{-1}\| \lesssim 1$. Combining this with (3.11) we get the thesis. □

Remark 3.5. Remark that when measuring with the Euclidean norm the vectors appearing in the linear system $\check{R}\underline{\check{u}} = g$, we are actually measuring the corresponding functions in the norm of the functional spaces where they naturally belong.

Theorem 3.4 states that if we express the stiffness matrix using the wavelet basis, we have an optimal diagonal preconditioner. However working directly with the wavelet basis has several drawbacks, namely that the matrix R is less sparse and more difficult and expensive to compute than the stiffness matrix A with respect to the scaling function basis. It is however possible to write down the preconditioned wavelet method in terms of this last matrix. We recall in fact that we have a (fast) algorithm (the *fast wavelet transform*) allowing us to compute the wavelet coefficients from the nodal coefficients of any function in V_j. Let us then denote by F the linear transformation corresponding to such algorithm. The following identity holds:

$$R = F^{-T} A F^{-1}.$$

We immediately obtain a preconditioner for A:

$$\kappa(D^{-1} F^{-T} A F^{-1} D^{-1}) \lesssim 1$$

uniformly in j.

This is the form that should be used in the implementation. Remark that the matrix $D^{-1}F^{-T}AF^{-1}D^{-1}$ is never assembled. Its action on a given vector is rather computed by applying sequentially the matrices D^{-1}, F^{-1}, A, F^{-T} and D^{-1}. The total number of operations is kept low thanks to different facts:

- the matrix A is sparse and then computing its action on a vector is cheap;
- the action of both F^{-1} and its transpose on any given vector can be computed by means of the inverse fast wavelet transform (which is also a fast algorithm).

The cost of applying the preconditioner basically reduces to the cost of running the fast wavelet transform algorithm.

3.2 Nonlinear Wavelet Methods for the Solution of PDEs

Perhaps the most significant contribution coming from the wavelet framework to the problem of numerically solving partial differential equations is the one related to the field of nonlinear approximation. Adaptive methods have an important place in the effort to efficiently tackle real life problems, and wavelets and the related concept of nonlinear wavelet approximation ([29]) brought a new point of view in this area that had also an impact on more classical methods (see, e.g., [34]).

3.2.1 Nonlinear vs. Linear Wavelet Approximation

Let us consider the problem of approximating a given function $f \in L^2(\Omega)$, $\Omega \subseteq \mathbb{R}^d$ a domain in $\mathbb{R}^d$, with N degrees of freedom (that is with a function which we can identify with N scalar coefficients). We distinguish between two approaches:

The first approach is the usual *linear approximation*: a space V_h of dimension N is fixed a priori and the approximation f_h is the $L^2(\Omega)$-projection of f on V_h (example: finite elements on a uniform grid with mesh size h related to N, in which case in dimension d the mesh size and the number of degrees of freedom verify $h \sim N^{-1/d}$).

It is well known that the behaviour of the linear approximation error is generally linked to the Sobolev regularity of f. In particular we cannot hope for a high rate of convergence if f has poor smoothness. Several remedies are available in this last case, like for instance adaptive approximation by performing a mesh refinement around the singularities of f. We have then the *nonlinear approximation* approach: a class of spaces X is chosen a priori. We then choose a space $V_N(f)$ of dimension N in X well suited to f. The approximation f_h is finally computed as an L^2-"projection" of f onto $V_N(f)$ (example: finite elements with N free nodes). In other words, we look for an approximation to f in the *nonlinear* space

$$\Sigma_N = \bigcup_{V_N \in X} V_N.$$

Three questions are of interest:

- What is the relation between the performance of nonlinear approximation and some kind of smoothness of the function to be approximated?
- How do we compute the nonlinear approximation of a given function f?
- How do we compute the nonlinear approximation of an unknown function u (solution of a PDE)?

In order to give answers to these three questions when we put ourselves in the framework of wavelet approximation let us see how the two approaches translate in such framework.

Linear Wavelet Approximation

Let the space V_h ($h = 2^{-j}$) be defined as $V_h = V_j$. The best L^2-approximation of f in V_h is its L^2-orthogonal projection $\Pi_h f$. As far as the approximation error is concerned $f \in H^s(\Omega)$ implies ($h \sim 2^{-j} \sim N^{-1/d}$)

$$\|f - \Pi_h f\|_{L^2(\Omega)} \leq \|f - P_j f\|_{L^2(\Omega)} \lesssim h^s \|f\|_{H^s(\Omega)}.$$

Nonlinear Wavelet Approximation

The idea is to look for an approximation to f in the nonlinear space

$$\Sigma_N = \Big\{ u = \sum_\lambda u_\lambda \psi_\lambda : \ \#\{\lambda : u_\lambda \neq 0\} \leq N \Big\}.$$

In order to construct an approximation to f in Σ_N define a *nonlinear projection* ([29]).

Definition 3.6. Let $f = \sum_\lambda f_\lambda \psi_\lambda$. The nonlinear projector $P_N : L^2 \to \Sigma_N$ is defined as

$$P_N f := \sum_{n=1}^{N} f_{\lambda_n} \psi_{\lambda_n}$$

where

$$|f_{\lambda_1}| \geq |f_{\lambda_2}| \geq |f_{\lambda_3}| \geq \dots |f_{\lambda_n}| \geq |f_{\lambda_{n+1}}| \geq \cdots$$

is a decreasing reordering of the wavelet coefficients.

By abuse of notation we will also denote by $P_N : \ell^2 \to \ell^2$ the operator mapping the sequence of coefficients of a function f to the sequence of coefficients of $P_N f$.

Remark 3.7. If the wavelet basis is orthonormal $\|f\|_{L^2(\Omega)} = \|(f_\lambda)\|_{\ell^2}$, then $P_N f$ is the best N-term approximation (the norm of the error is the ℓ^2-norm of the sequence of discarded coefficients). In any case, since $\|f\|_{L^2(\Omega)} \simeq \|(f_\lambda)\|_{\ell^2}$, we have that $P_N f$ is the best approximation in an L^2-equivalent norm.

Let us now see which properties of f guarantee that

$$\|f - P_N f\|_{L^2(\Omega)} \lesssim h^r \sim N^{-r/d}.$$

We have the following theorem ([29], see also [16]).

Theorem 3.8. *$f \in B^{s,q}_q(\Omega)$ with $s > 0$ and $q : d/q = d/2 + s$ implies*

$$\|f - P_N f\|_{L^2(\Omega)} \lesssim \|f\|_{B^{s,q}_q(\Omega)} N^{-s/d}.$$

Proof. Recall the characterization of the Besov norm in terms of the wavelet coefficients: with the choice $q : d/q = d/2 + s$ we have that

$$\Big\| \sum_\lambda f_\lambda \psi_\lambda \Big\|_{B^{s,q}_q(\Omega)} \simeq \|(f_\lambda)_\lambda\|_{\ell^q}.$$

Let us consider the decreasing reordering of the coefficients:

$$|f_{\lambda_1}| \geq |f_{\lambda_2}| \geq |f_{\lambda_3}| \geq \dots |f_{\lambda_n}| \geq |f_{\lambda_{n+1}}| \geq \cdots.$$

We can easily see that

$$n|f_{\lambda_n}|^q \leq \sum_{k\leq n} |f_{\lambda_k}|^q \leq \sum_{\lambda} |f_\lambda|^q \leq \|f\|^q_{B^{s,q}_q(\Omega)},$$

that is

$$|f_{\lambda_n}| \leq n^{-1/q} \|f\|_{B^{s,q}_q(\Omega)}.$$

Now we can write

$$\|f - P_N f\|_{L^2(\Omega)} = \Big\| \sum_{n>N} f_{\lambda_n} \psi_{\lambda_n} \Big\|_{L^2(\Omega)} \simeq \left(\sum_{n>N} |f_{\lambda_n}|^2 \right)^{1/2}$$

$$\lesssim \|f\|_{B^{s,q}_q(\Omega)} \left(\sum_{n>N} n^{-2/q} \right)^{1/2} \lesssim \|f\|_{B^{s,q}_q(\Omega)} N^{-1/q+1/2}. \qquad \square$$

Remark 3.9. Remark that, for $q < 2$, the space $B^{s,q}_q(\Omega) \supset H^s(\Omega)$. In particular there exists a wide class of functions which are not in $H^s(\Omega)$ but that belong to $B^{s,q}_q(\Omega)$. For such functions, nonlinear approximation will be of order h^s, while linear approximation will go to the order of approximation allowed by the (lower) Sobolev regularity.

Nonlinear Approximation in H^s

If we want to approximate f in $H^s(\Omega)$, rather than in $L^2(\Omega)$, the idea is to rescale the basis functions so that $f = \sum \check{f}_\lambda \check{\psi}_\lambda$ with $\|f\|_{H^s(\Omega)} = \|(\check{f}_\lambda)_\lambda\|_{\ell^2}$. Remark that for $q : d/q = d/2 + r$

$$\|f\|_{B^{s+r,q}_q(\Omega)} \simeq \|(\check{f}_\lambda)_\lambda\|_{\ell^q}.$$

We then apply the same procedure to the sequence $(\check{f}_\lambda)_\lambda$. In particular we define this time the nonlinear projector as

Definition 3.10. Let $f = \sum_\lambda \check{f}_\lambda \check{\psi}_\lambda$. The nonlinear projector $P_N : H^s(\Omega) \to \Sigma_N$ is defined as

$$P_N f := \sum_{n=1}^{N} \check{f}_{\lambda_n} \check{\psi}_{\lambda_n}$$

where

$$|\check{f}_{\lambda_1}| \geq |\check{f}_{\lambda_2}| \geq |\check{f}_{\lambda_3}| \geq \dots |\check{f}_{\lambda_n}| \geq |\check{f}_{\lambda_{n+1}}| \geq \dots$$

is a decreasing reordering of the wavelet coefficients.

We have the following theorem, whose proof is identical to the proof of Theorem 3.8

Theorem 3.11. $f \in B^{s+r,q}_q$ *implies*

$$\|f - P_N f\|_{H^s(\Omega)} \lesssim \|f\|_{B^{s+r,q}_q(\Omega)} N^{-r/d}.$$

3.2.2 Nonlinear Solution of PDEs

Depending on different factors, like the regularity of the data and of the domain Ω, the solution of a differential problem may be smooth, or it may present some singularity. In the last case, the fact that using some adaptive technique – in which the approximating space is tailored to the function u itself – is necessary in order to get a good approximation rate is well accepted. The considerations in the above section allow to rigorously formalize such fact and provide, in the wavelet context, a simple and efficient strategy for adaptively approximating u, if this was given. However, in the partial differential equations framework the function that one needs to approximate is not known. Standard adaptive methods are based on an iterative procedure:

- given an approximation space, compute an approximation to the solution of the problem within the given space;
- looking at the computed approximation, design a new approximation space (using, for instance, some kind of error estimator);
- iterate until the computed solution is satisfactory.

Nonlinear wavelet solution of PDEs is based on a different approach. To fix the ideas let us consider a simple example, namely the *reaction-diffusion* equation: given f find u such that

$$-\Delta u + u = f \qquad \text{in } \Omega, \qquad \Omega \subseteq \mathbb{R}^d. \tag{3.12}$$

The idea is to first re-write the problem as an "infinite-dimensional discrete" problem as follows: letting $u = \sum_\lambda \check{u}_\lambda \check{\psi}_\lambda$ be the unknown solution ($\underline{u} = (u_\lambda)_\lambda \in \ell^2$ being the corresponding infinite unknown vector), test the equation against the infinite set of test functions $v_h = \check{\psi}_\mu$. The PDE becomes

$$\check{\mathcal{R}}\underline{u} = \underline{f}$$

where $\check{\mathcal{R}} = (r_{\lambda,\mu})$ is the bi-infinite stiffness matrix and $\underline{f} = \{f_\mu\}$ the bi-infinite right-hand side vector, with

$$r_{\lambda,\mu} = a(\check{\psi}_\mu, \check{\psi}_\lambda) = \int_\Omega (\nabla\check{\psi}_\mu \cdot \nabla\check{\psi}_\lambda + \check{\psi}_\mu\check{\psi}_\lambda), \qquad f_\lambda = \langle f, \check{\psi}_\lambda \rangle.$$

For such a bi-infinite linear system we can formally write down an iterative solution scheme. To fix the ideas let us consider a simple Richardson scheme:

The Richardson Scheme for the Continuous Problem

- `initial guess` $\underline{u}^0 = 0$
- $\underline{u}^n \longrightarrow \underline{u}^{n+1}$

– `compute` $r^n_\lambda = f_\lambda - (\check{\mathcal{R}}\underline{u}^n)_\lambda$

– $\underline{u}^{n+1} = \underline{u}^n + \theta \underline{r}^n$

- `iterate until error` $\le$ `tolerance.`

It is well known that the behavior of the above scheme is related to the behavior of the operator $I - \theta\check{\mathcal{R}}$. In this regard we have the following lemma ([18]).

Lemma 3.12. *There exist two constants $\tilde{q}$ and θ_0, such that for all θ with $0 < \theta \le \theta_0$ and for all q with $\tilde{q} < q \le 2$,*

$$\|I - \theta\check{\mathcal{R}}\|_{\mathcal{L}(\ell^q,\ell^q)} \le \rho < 1. \tag{3.13}$$

Thanks to such a lemma (which is closely related to the preconditioning results described in the previous section) it is not difficult to prove that there exists a θ_0 such that the Richardson scheme for the continuous problem converges provided $\theta < \theta_0$.

The Nonlinear Richardson Scheme

The idea is to modify the above scheme by forcing the functions $u^n = \sum_\lambda \check{u}_\lambda \check{\psi}_\lambda$ to be in the nonlinear space Σ_N. This reduces to forcing the iterates $\underline{u}^n$ to have at most N nonzero entries. In order to do so we plug into the scheme the nonlinear projector P_N of Definition 3.10 as follows:

- `initial guess` $\underline{u}^0 = 0$
- $\underline{u}^n \longrightarrow \underline{u}^{n+1}$

  – `compute` $r^n_\lambda = f_\lambda - (\check{\mathcal{R}}\underline{u}^n)_\lambda$

  – $\underline{u}^{n+1} = P_N(\underline{u}^n + \theta\underline{r}^n) \qquad (u^{n+1} = \sum_\lambda u^{n+1}_\lambda \psi_\lambda \in \Sigma_N)$

- `iterate until error` $\le$ `tolerance.`

By construction the result of this procedure will belong to Σ_N. This scheme is however not yet computable since it involves operations on infinite matrices and vectors. We will give an idea on how this issue can be dealt with in Section 3.5.

Theorem 3.13. *If $u \in B^{s+t,q}_q(\Omega)$ with $\tilde{q} < q \le 2$, $t = d/q - d/2$, then there exists a θ_0 s.t. for $0 < \theta \le \theta_0$ it holds:*

- *stability: we have*

$$\|\underline{u}^n\|_{\ell^2} \lesssim \|\underline{f}\|_{\ell^2} + \|\underline{u}^0\|_{\ell^2}, \quad \forall n \in \mathbb{N}.$$

- *approximation error estimate: for $\underline{e}^n = \underline{u}^n - \underline{u}$ it holds:*

$$\|\underline{e}^n\|_{\ell^2} \leq \rho^n \|\underline{e}^0\|_{\ell^2} + \frac{C}{1-\rho} N^{-t/d},$$

where C is a constant depending only on the initial data.

Proof. Stability. We have, using the ℓ^2-boundedness of P_N as well as (3.13),

$$\begin{aligned}\|\underline{u}^n\|_{\ell^2} &= \|P_N(\underline{u}^{n-1} + \theta(\underline{g} - \check{\mathcal{R}}\underline{u}^{n-1}))\|_{\ell^2} \leq \|(1 - \theta\check{\mathcal{R}})\underline{u}^{n-1} + \theta\underline{g}\|_{\ell^2} \\ &\leq \|\theta\underline{g}\|_{\ell^2} + \rho\|\underline{u}^{n-1}\|_{\ell^2}.\end{aligned}$$

Iterating this bound for n decreasing to 0 we obtain

$$\|\underline{u}^n\|_{\ell^2} \leq \left(\sum_{i=0}^{n-1} \rho^i\right) \|\theta\underline{g}\|_{\ell^2} + \rho^n\|\underline{u}^0\|_{\ell^2},$$

which gives us the stability bound. In the same way we can prove that

$$\|\underline{u}^n\|_{\ell^q} \lesssim \|\underline{f}\|_{\ell^q} + \|\underline{u}^0\|_{\ell^q}, \quad \forall n \in \mathbb{N}. \tag{3.14}$$

Convergence. We write down an error equation

$$\underline{e}^{n+1} = \underline{e}^n - \theta\check{\mathcal{R}}\underline{e}^n + \underline{\varepsilon}^n,$$

with

$$\underline{\varepsilon}^n = P_N(\underline{u}^n + \theta(\underline{f} - \check{\mathcal{R}}\underline{u}^n)) - (\underline{u}^n + \theta(\underline{f} - \check{\mathcal{R}}\underline{u}^n)).$$

We take the ℓ^2-norm, and, using (3.13) once again,

$$\begin{aligned}\|\underline{e}^{n+1}\|_{\ell^2} &\leq \rho\|\underline{e}^n\|_{\ell^2} + \|\underline{\varepsilon}^n\|_{\ell^2} \leq \sum_{i=0}^{n} \rho^{n-i}\|\underline{\varepsilon}^i\|_{\ell^2} + \rho^{n+1}\|e_0\|_{\ell^2} \\ &\leq \left(\max_{0\leq k\leq n} \|\underline{\varepsilon}^k\|_{\ell^2}\right) \sum_{k=1}^{n} \rho^k + \rho^{n+1}\|e_0\|_{\ell^2}.\end{aligned}$$

Let us bound ε^n. Using (3.14) we have

$$\|\underline{\varepsilon}^n\|_{\ell^2} \leq N^{-t/d}\|\underline{u}^n + \theta(\underline{f} - \check{\mathcal{R}}\underline{u}^n)\|_{\ell^q} \leq N^{-t/d}\|(I - \theta\check{\mathcal{R}})\underline{u}^n + \theta\underline{f}\|_{\ell^q} \lesssim N^{-t/d}. \qquad \square$$

3.3 Wavelet Stabilisation of Unstable Problems

As we saw in the previous chapter, wavelet bases give us a way of practically realizing equivalent scalar products for Sobolev spaces of negative and/or fractionary index. This is the key ingredient of *wavelet stabilisation.*

To fix the ideas, let us consider a simple model problem, the Stokes equation, though in general the ideas that we are going to present can be easily generalised to a wide class of differential equations (provided a wavelet basis for the domain of definition of the problem can be constructed). Given $f \in (H^{-1}(\Omega))^d$ (Ω bounded domain of $\mathbb{R}^d$, $d = 2, 3$) find $u : \Omega \longrightarrow \mathbb{R}^d$ and $p : \Omega \longrightarrow \mathbb{R}$ such that

$$\begin{cases} -\Delta u + \nabla p &= f, \\ \nabla \cdot u &= 0, \end{cases} \tag{3.15}$$

$$u = 0, \text{ on } \partial\Omega, \qquad \int_\Omega p = 0. \tag{3.16}$$

or, in variational formulation,

Problem 3.14 (Stokes Problem). Find $u \in U = (H_0^1(\Omega))^d$ and $p \in Q = L_0^2(\Omega)$ such that for all $v \in U$ and $q \in Q$ one has

$$\int_\Omega (\nabla u \cdot \nabla v - p \nabla \cdot v) = \int_\Omega f \cdot v, \tag{3.17}$$

$$\int_\Omega \nabla \cdot uq = 0, \tag{3.18}$$

where $L_0^2(\Omega) = \{q \in L^2(\Omega) : \int_\Omega q = 0\} \subset L^2(\Omega)$ denotes the space of L^2-functions with zero mean value. It is well known that the bilinear form $a : (U \times Q) \times (U \times Q) \longrightarrow \mathbb{R}$,

$$a(u, p; v, q) = \int_\Omega \nabla u \cdot \nabla v - \int_\Omega \nabla \cdot vp + \int_\Omega \nabla \cdot uq, \tag{3.19}$$

corresponding to such a problem is not coercive. Existence and uniqueness of the solution of such a problem are ensured by the *inf-sup* condition

$$\inf_{q \in Q} \sup_{v \in U} \frac{\int_\Omega \nabla \cdot vq}{\|v\|_{H^1(\Omega)} \|q\|_{L^2(\Omega)}} \geq \alpha > 0. \tag{3.20}$$

As a consequence, in solving such a problem, an arbitrary choice of the discretisation spaces for the velocity u and for the pressure p can lead to an unstable discrete problem. In order to have stable discretisations, the velocity and pressure approximation spaces U_h and Q_h need to be coupled in such a way that they satisfy a *discrete inf-sup condition* [13]:

$$\inf_{q_h \in Q_h} \sup_{v_h \in U_h} \frac{\int_\Omega \nabla \cdot v_h q_h}{\|v_h\|_{H^1(\Omega)} \|q_h\|_{L^2(\Omega)}} \geq \alpha_1 > 0, \tag{3.21}$$

with α_1 independent of the discretisation step h. Many pairs of discretization spaces are available satisfying such a property. However, for several different reasons, it might be desirable to be able to design methods based on discretization spaces that do not satisfy it. *Stabilised methods* are a class of method that, through different means, give the possibility of effectively avoiding the instabilities deriving from the lack of validity of the inf-sup condition.

The idea of the *wavelet stabilised method* [3] is to introduce the following equivalent formulation of the Stokes problem.

Problem 3.15 (Regularized Stokes problem). Find $u \in (H_0^1(\Omega))^d$ and $p \in L_0^2(\Omega)$ such that for all $v \in (H_0^1(\Omega))^d$ and $q \in L_0^2(\Omega)$ we have

$$\int_\Omega (\nabla u \cdot \nabla v - p\nabla \cdot v) = \int_\Omega f \cdot v, \tag{3.22}$$

$$\int_\Omega \nabla \cdot uq + \gamma(-\Delta u + \nabla p, \nabla q)_{-1} = \gamma(f, \nabla q)_{-1}, \tag{3.23}$$

where $\gamma > 0$ is a mesh-independent constant to be chosen and where $(\cdot,\cdot)_{-1} : (H^{-1}(\Omega))^d \times (H^{-1}(\Omega))^d \longrightarrow \mathbb{R}$ is the equivalent scalar product for the space $(H^{-1}(\Omega))^d$ defined according to (2.49). It is easy to check that the bilinear form $a_{\text{stab}} : ((H_0^1(\Omega))^d \times L_0^2(\Omega))^2 \longrightarrow \mathbb{R}$ which is defined by

$$a_{\text{stab}}(u, p; v, q) = \int_\Omega \nabla u \cdot \nabla v - \int_\Omega p\nabla \cdot v + \int_\Omega \nabla \cdot uq + \gamma(-\Delta u + \nabla p, \nabla q)_{-1},$$

and which corresponds to such a formulation, is continuous. Moreover it is possible to prove that it is coercive for suitable choices of the constant γ. More precisely the following lemma holds [2].

Lemma 3.16. *There exists a constant γ_0 (depending on the domain Ω) such that if γ satisfies $0 < \gamma < \gamma_0$, then the bilinear form a_{stab} is coercive.*

Proof. We have

$$a_{\text{stab}}(u, p; u, p) \geq |u|^2_{H^1(\Omega)} + \gamma\|\nabla p\|^2_{H^{-1}(\Omega)} - \gamma\|\Delta u\|_{H^{-1}(\Omega)}\|\nabla p\|_{H^{-1}(\Omega)}.$$

We now observe that, using the Poincaré inequality, we can write, for $\kappa > 0$,

$$\|\Delta u\|_{H^{-1}(\Omega)}\|\nabla p\|_{H^{-1}(\Omega)} \leq C|u|_{H^1(\Omega)}\|\nabla p\|_{H^{-1}(\Omega)} \lesssim \frac{C\kappa}{2}|u|^2_{H^1(\Omega)} + \frac{C}{2\kappa}\|\nabla p\|^2_{H^{-1}(\Omega)}.$$

This yields

$$a_{\text{stab}}(u, p; u, p) \geq \left(1 - \frac{1}{2}\gamma C\kappa\right)|u|^2_{H^1(\Omega)} + \gamma\left(1 - \frac{C}{2\kappa}\right)\|\nabla p\|^2_{H^{-1}(\Omega)}.$$

By choosing $\kappa = C$ and $\gamma_0 = 2/C^2$ (so that for $\gamma < \gamma_0$ we have $(1 - \frac{1}{2}\gamma C\kappa) > 0$) we obtain that

$$a_{\text{stab}}(u, p; u, p) \gtrsim |u|^2_{H^1(\Omega)} + \gamma\|\nabla p\|^2_{H^{-1}(\Omega)}$$

(the constant in the inequality depending on γ). We now observe that, since the Stokes operator is boundedly invertible, we can write

$$\|u\|^2_{H^1(\Omega)} + \|p\|^2_{L^2(\Omega)} \lesssim \|-\Delta u + \nabla p\|^2_{H^{-1}(\Omega)} + \|\nabla \cdot u\|^2_{L^2(\Omega)} \lesssim |u|^2_{H^1(\Omega)} + \|\nabla p\|^2_{H^{-1}(\Omega)},$$

which allows us to conclude. □

Given any finite-dimensional subspaces $U_h \subset (H_0^1(\Omega))^d$ and $Q_h \subset L_0^2(\Omega)$ we consider the following discrete problem.

Problem 3.17 (Discrete Stabilised Problem). Find $u_h \in U_h$ and $p_h \in Q_h$ such that for all $v_h \in U_h$ and $q_h \in Q_h$ we have

$$\int_\Omega (\nabla u_h \cdot \nabla v_h - p_h \nabla \cdot v_h) = \int_\Omega f \cdot v_h, \tag{3.24}$$

$$\int_\Omega \nabla \cdot u_h q_h + \gamma(-\Delta u_h + \nabla p_h, \nabla q_h)_{-1} = \gamma(f, \nabla q_h)_{-1}. \tag{3.25}$$

Using the standard theory for the Galerkin discretisation of coercive operators ([15]) we immediately obtain the following error estimate.

Proposition 3.18. *Let* (u, p) *be the solution of problem* (3.17) *and* (u_h, p_h) *the solution of problem* (3.24) *Then the following error estimate holds:*

$$\|u-u_h\|_{H^1(\Omega)}+\|p-p_h\|_{L^2(\Omega)} \lesssim \big(\inf_{v_h \in U_h} \|u-v_h\|_{H^1(\Omega)}+ \inf_{q_h \in Q_h} \|p-q_h\|_{L^2(\Omega)}\big). \tag{3.26}$$

The use of the stabilised formulation gives then rise to an optimal discretisation of the Stokes problem, for which the choice of the approximation spaces is not subject to limitations on the coupling of the discretisations for velocity and pressure. Remark that the stabilized problem falls in the framework considered in the previous sections. We can then apply for its solution all the techniques described therein.

Many problems share the same characteristics as the Stokes problem, and can therefore benefit from an analogous approach. An abstract result can be found in [2] and [5]. This has been applied in the domain decomposition framework (see [9], [6]), as well as to the Lagrange multiplier formulation of Dirichlet problems [7]. The use of wavelets to realize negative and/or fractionary scalar products has also been applied in the framework of the *Least Squares* method (see [12],[24]). An analogous technique, based on the wavelet evaluation of a $-1/2$ scalar product, has also been applied to *convection-diffusion* problems with dominating convection [8].

3.4 A-Posteriori Error Estimates

The norm equivalence for the Sobolev spaces with negative index can be exploited also for designing *a posteriori* error estimators for the numerical solution of partial differential equations. Again, to fix the ideas, let us consider the reaction-diffusion equation (3.12). Assume that we have somehow computed an approximation u_h to the solution u, with u_h in some finite-dimensional approximation space. The method used to obtain such an approximation plays no essential role in the following considerations (though some simplifications can take place if the numerical solution is obtained using the Galerkin method with the same wavelet basis functions used to characterize the $(H^1(\Omega))'$ norm in the following). It is well known that the *reaction-diffusion* operator $-\Delta+1$ is an isomorphism between $H^1(\Omega)$ and its dual $(H^1(\Omega))'$. Then we can bound

$$\|u-u_h\|_{H^1(\Omega)} \simeq \|(1-\Delta)(u-u_h)\|_{(H^1(\Omega))'} = \|f+(\Delta-1)u_h\|_{(H^1(\Omega))'}.$$

Once again, the norm equivalence in terms of wavelet coefficients provides us a practical way of computing the (equivalent) $(H^1(\Omega))'$ norm on the right-hand side (which is otherwise usually replaced by some weighted $L^2(\Omega)$-norm). By using wavelet bases we can in fact replace the $(H^1(\Omega))'$-norm by its equivalent in terms of wavelet coefficients, obtaining the bound

$$\|u-u_h\|^2_{H^1(\Omega)} \simeq \sum_j \sum_{\lambda\in\Lambda_j} 2^{-2j} |\langle f+(\Delta-1)u_h, \tilde{\psi}_\lambda\rangle|^2. \tag{3.27}$$

The term $e_\lambda = |\langle f+\Delta u_h, \tilde{\psi}_\lambda\rangle|$ plays then the role of an error indicator, that for $\lambda = (j,k)$ gives an information on the error at position $k/2^j$ and frequency $\sim 2^j$. Remark that the sum on the right-hand side of (3.27) is infinite. In order for the above error indicator to be applicable one will have to find a way of truncating it to a finite sum while still keeping the validity of an estimate of the form (3.27). This can be done by an argument similar to the one described in the next section.

3.5 Operations on Infinite Matrices and Vectors

For most of the applications considered we ended up with methods that implied the computation of either an infinite sum or of an infinite matrix/vector multiplication involving

- matrices $\mathcal{A}$ expressing a differential operator with good properties (continuity, coercivity, ...);
- vectors $\vec{u}$ of wavelet coefficients of a discrete function.

This is the case of both the computation of the equivalent scalar product $(\cdot,\cdot)_{-1}$ appearing in Section 3.3 and of the infinite matrix/vector multiplication appearing in Section 3.2. These matrices and vectors are not directly *maniable*. However, thanks to the properties of wavelets it is in general possible to replace the infinite sum by a finite one without substantially changing the resulting method.

For the sake of simplicity let us concentrate on the case of Ω a bounded domain, so that for any fixed level j the cardinality of Λ_j is finite. To fix the ideas, let us consider the equivalent scalar product $(\cdot,\cdot)_{-1}$ in Section 3.3. Heuristically, the argument that we have in mind is that *if a discrete function satisfies an inverse inequality ($\sim$ it is "low frequency"), then the levels in the infinite sum corresponding to "high frequency" components will be negligible and then the infinite sum* *in* (2.49) *can be truncated.* Just for this example we would like to show how this heuristics can be made rigorous. The aim is to replace the scalar product

$$(F,G)_{-1} = \sum_j \sum_{\lambda\in\Lambda_j} 2^{-2j} \langle F, \tilde{\psi}_\lambda\rangle \langle G, \tilde{\psi}_\lambda\rangle$$

in (3.22) with a computable bilinear form of the form

$$(F,G)_{-1,J} = \sum_{j\le J} \sum_{\lambda\in\Lambda_j} 2^{-2j} \langle F, \tilde{\psi}_\lambda\rangle \langle G, \tilde{\psi}_\lambda\rangle,$$

while retaining the properties (coercivity on the discrete space of the stabilized operator) of the resulting discrete method.

Since the aim of adding the stabilisation term to the original equation is to obtain control on the pressure p_h through a coercivity argument, and since the velocity is controlled through coercivity already for the original bilinear form a, the fundamental property of $(\cdot,\cdot)_{-1}$ that we want to keep, in this case, is that for arbitrary elements $q_h \in Q_h$ one has

$$(\nabla q_h, \nabla q_h)_{-1,J} = \sum_{j\le J} \sum_{\lambda\in\Lambda_j} 2^{-2j} |\langle \nabla q_h, \tilde{\psi}_\lambda\rangle|^2 \gtrsim \|\nabla p_h\|_{H^{-1}(\Omega)}. \tag{3.28}$$

We will have to replace the heuristical concept q_h *is "low frequency"* by a suitable inverse inequality: more precisely we will assume that $\nabla Q_h \subset H^t(\Omega)$ for some t with $-1 < t$, and that for all $q_h \in Q_h$

$$\|\nabla q_h\|_t \lesssim h^{-t-1} \|\nabla q_h\|_{-1}.$$

Under this assumption it is actually possible to prove that there exists a $J = J(h)$ depending on the mesh-size parameter h such that (3.28) holds. The proof is simple and gives an idea of how this kind of argument works in general. For any given $J > 0$ we can write

$$\|\nabla q_h\|^2_{H^{-1}(\Omega)} \simeq \sum_{j\le J} \sum_{\lambda\in\Lambda_j} 2^{-2j} |\langle \nabla q_h, \tilde{\psi}_\lambda\rangle|^2 + \sum_{j>J} \sum_{\lambda\in\Lambda_j} 2^{-2j} |\langle \nabla q_h, \tilde{\psi}_\lambda\rangle|^2.$$

Let us analyse the last term:

$$\begin{aligned}\sum_{j>J} \sum_{\lambda\in\Lambda_j} 2^{-2j} |\langle \nabla q_h, \tilde{\psi}_\lambda\rangle|^2 &\lesssim 2^{-2(t+1)J} \sum_{j>J} \sum_{\lambda\in\Lambda_j} 2^{2tj} |\langle \nabla q_h, \tilde{\psi}_\lambda\rangle|^2 \\ &\lesssim 2^{-2(t+1)J} \|\nabla q_h\|^2_{H^t(\Omega)} \\ &\lesssim 2^{-2(t+1)J} h^{-2(t+1)} \|\nabla q_h\|^2_{H^{-1}(\Omega)}.\end{aligned}$$

Then we can write

$$\|\nabla q_h\|^2_{H^{-1}(\Omega)} \le \sum_{j\le J} \sum_{\lambda} 2^{-2j} |\langle \nabla q_h, \tilde{\psi}_\lambda\rangle|^2 + C' 2^{-2(t+1)J} h^{-2(t+1)} \|\nabla q_h\|^2_{H^{-1}(\Omega)}$$

whence

$$(1 - C'2^{-2(t+1)J}h^{-2(t+1)})\|\nabla q_h\|^2_{H^{-1}(\Omega)} \lesssim \sum_{j\leq J}\sum_{\lambda} 2^{-2j}|\langle \nabla q_h, \tilde{\psi}_\lambda\rangle|^2$$

which, provided J is chosen in such a way that $(1 - C'2^{-2(t+1)J}h^{-2(t+1)}) \leq 1/2$, yields

$$\|\nabla q_h\|^2_{H^{-1}(\Omega)} \lesssim \sum_{j\leq J}\sum_{\lambda} 2^{-2j}|\langle \nabla q_h, \tilde{\psi}_\lambda\rangle|^2.$$

It is not difficult to show that Lemma 3.16 still holds if we replace $(\cdot,\cdot)_{-1}$ by $(\cdot,\cdot)_{-1,J}$ in the definition of the stabilized method (3.22), (3.23). In doing so, we obtain a computable method with the same characteristics. In particular continuity and coercivity of the stabilized bilinear form will still hold, yielding optimal error estimates.

An analogous result holds for the evaluation of the infinite matrix/vector multiplication $\mathcal{A}\vec{u}$ that is found in the nonlinear Richardson scheme of Chapter 3.2. Taking advantage of the fact that, by construction, $\vec{u}$ is the linear combination of a finite number of wavelets and that, thanks to the localisation property of wavelets, the infinite matrix $\mathcal{A}$ can be sparsified (by retaining, per line, only a finite number of entries which are, in absolute value, greater than a suitable tolerance ε) one can compute the products $\mathcal{A}\vec{u}$ in a finite number of operations within a given tolerance [18].

Bibliography

[1] N. Hall, B. Jawerth Andersson, L. and G. Peters, Wavelets on closed subsets of the real line. Technical report, 1993.

[2] C. Baiocchi and F. Brezzi, Stabilization of unstable methods. In P.E. Ricci, editor, *Problemi attuali dell'Analisi e della Fisica Matematica*. Università "La Sapienza", Roma, 1993.

[3] S. Bertoluzza, Wavelets for the numerical solution of the stokes equation. Technical report, 2007. 2007.

[4] S. Bertoluzza, Recent Developments in Wavelet Methods for the solution of PDE's. in *Atti del XVII Congresso dell'Unione Matematica Italiana – Milano, 8–13 Settembre 2003*, UMI (Bologna), 2004.

[5] S. Bertoluzza, Stabilization by multiscale decomposition. *Appl. Math. Lett.* **11** no. 6 (1998), 129–134.

[6] S. Bertoluzza, Analysis of a stabilized three fields domain decomposition method. Technical Report 1175, I.A.N.-C.N.R., 2000. Numer. Math., to appear.

[7] S. Bertoluzza, Wavelet stabilization of the Lagrange multiplier method. *Numer. Math.* **86** (2000), 1–28.

[8] S. Bertoluzza, C. Canuto, and A. Tabacco, Stable discretization of convection-diffusion problems via computable negative order inner products. *SINUM* **38** (2000), 1034–1055.

[9] S. Bertoluzza and A. Kunoth, Wavelet stabilization and preconditioning for domain decomposition. *I.M.A. Jour. Numer. Anal.* **20** (2000), 533–559.

[10] S. Bertoluzza and M. Verani, Convergence of a non-linear wavelet algorithm for the solution of PDE's. *Appl. Math. Lett.* **16** no. 1 (2003).

[11] S. Bertoluzza, S. Mazet and M. Verani, A nonlinear Richardson algorithm for the solution of elliptic PDE's, *M3AS* **13** (2003), 143–158.

[12] J.H. Bramble, R.D. Lazarov, and J.E. Pasciak, A least square approach based on a discrete minus one product for first order systems. Technical Report BNL-60624, Brookhaven National Laboratory, 1994.

[13] F. Brezzi and M. Fortin, *Mixed and Hybrid Finite Element Methods.* Springer, 1991.

[14] C. Canuto, A. Tabacco, and K. Urban, The wavelet element method. II. Realization and additional features in 2D and 3D. *Appl. Comput. Harmon. Anal.* **8** (2000).

[15] P.G. Ciarlet, *The Finite Element Method for Elliptic Problems.* North-Holland, Amsterdam, 1976.

[16] A. Cohen, *Numerical Analysis of Wavelet Methods.* Elsevier, 2003.

[17] A. Cohen, W. Dahmen, and R. De Vore, Multiscale decomposition on bounded domains. Preprint.

[18] A. Cohen, W. Dahmen, and R. De Vore, Adaptive wavelet methods for elliption operator equations – convergence rates. Technical report, IGPM – RWTH-Aachen, 1998. To appear in Math. Comp.

[19] A. Cohen, I. Daubechies and J. Feauveau, Biorthogonal Bases of compactly supported Wavelets, *Comm. Pure Appl. Math.* **45** (1992), 485–560.

[20] A. Cohen, I. Daubechies, and P. Vial, Wavelets on the interval and fast wavelet transforms. *ACHA* **1** (1993), 54–81.

[21] A. Cohen and R. Masson, Wavelet adaptive method for second order elliptic problems: boundary conditions and domain decomposition. *Numer. Math.* **86** no. 2 (2000), 193–238.

[22] W. Dahmen, Stability of multiscale transformations. *Journal of Fourier Analysis and Applications* **2** no. 4 (1996), 341–361.

[23] W. Dahmen and A. Kunoth, Multilevel preconditioning. *Numerische Mathematik* **63** no. 3 (1992), 315–344.

[24] W. Dahmen, A. Kunoth, and R. Schneider, Wavelet least square methods for boundary value problems. *Siam J. Numer. Anal.* (2002).

[25] W. Dahmen and C.A. Michelli, Using the refinement equation for evaluating integrals of wavelets. *SIAM J. Numer. Anal.* **30** (1993), 507–537.

[26] W. Dahmen and R. Schneider, Composite wavelet bases for operator equations. *Math. Comp.* **68** no. 228 (1999), 1533–1567.

[27] W. Dahmen and R. Stevenson, Element-by-element construction of wavelets satisfying stability and moment monditions. *SIAM J. Numer. Anal.* **37** no. 1 (1999), 319–352.

[28] I. Daubechies, *Ten lectures on wavelets.* Society for Industrial and Applied Mathematics (SIAM), Philadelphia, PA, 1992.

[29] R.A. DeVore, Nonlinear approximation. *Acta Numerica* (1998).

[30] S. Jaffard, Wavelets and Analysis of Partial Differential Equations, *Proceedings du Séminaire de Mathématiques appliquées du Collège de France*, année 1990–1991.

[31] P.G. Lemarié, Fonctions à support compact dans les analyses multi-résolutions, *Rev. Mat. Iberoamericana* **7** no. 2 (1991), 157–182.

[32] Y. Maday, V. Perrier and J.C. Ravel, Adaptivité dynamique sur bases d'ondelettes pour l'approximation d'équations aux derivées partielles. *C. R. Acad. Sci Paris* **312**, (Série I) (1991), 405–410.

[33] Y. Meyer, Wavelets and operators. In Ingrid Daubechies, editor, *Different Perspectives on Wavelets*, volume 47 of *Proceedings of Symposia in Applied Mathematics*, pages 35–58. American Math. Soc., Providence, RI, 1993. From an American Math. Soc. short course, Jan. 11–12, 1993, San Antonio, TX.

[34] E. Bänsch, P. Morin and R.H. Nochetto, An Adaptive Uzawa FEM for the Stokes Problem: Convergence without the Inf-Sup Condition, *SIAM J. Numer. Anal.* **40** no. 4 (2002), 1207–1229.

[35] Y. Saad and H.A. van der Vorst, Iterative solution of linear systems in the 20th century, *J. Comput. Appl. Math.* **123** no. 1 (2000).

[36] J.R. Shewchuk, An Introduction to the Conjugate Gradient Method Without the Agonizing Pain, August 1994.
`http://www.cs.cmu.edu/~quake-papers/painless-conjugate-gradient.pdf`

[37] H. Triebel, *Interpolation Theory, Function Spaces, Differential Operators.* North Holland, 1978.

Part II

High-Order Shock-Capturing Schemes for Balance Laws

Giovanni Russo

Chapter 1

Introduction

The purpose of these lecture notes is to provide an overview of the modern finite-volume and finite-difference shock-capturing schemes for systems of conservation and balance laws.

An emphasis is put in trying to give a unified view of such schemes, by identifying the essential aspects in their construction, namely conservation form, numerical flux function, nonlinear reconstruction, for the space discretization, and high-order accuracy and stability for time discretization.

Also, we shall attempt to present some of the various forms in which the equations can be discretized, by comparing method-of-lines approach for unstaggered finite-difference and finite-volume with time discretizations more entangled with space discretizations.

After a brief introduction on hyperbolic systems of conservation laws, we describe some classical numerical scheme in conservation form. The mathematical theory of quasilinear hyperbolic systems of conservation laws is used as a guideline in the construction and development of shock-capturing schemes, that imports from it relevant concepts such as conservation form, entropy inequality, propagation along the characteristics, and so on.

The next section is devoted to the construction of high-order shock-capturing finite-volume schemes. The schemes are obtained in a semidiscrete form, by performing first space discretization, and then time discretization. The key role of the numerical flux function and nonlinear (essentially) non-oscillatory reconstruction will be emphasized. As an alternative to finite-volume discretization, high-order finite-difference schemes can be constructed, and the relative merits of finite-volume and finite-difference methods will be discussed.

The third part of the notes concerns the treatment of systems of balance laws. For standard problems, in which the source term is not stiff, the extension of the finite-difference or finite-volume schemes to systems with source is straightforward. There are, however, two cases in which such an extension requires a more detailed investigation.

One case concerns the systems with stiff source, where time discretization has to be chosen with care, in order to construct an efficient scheme. As we shall see, Runge-Kutta Implicit-Explicit schemes (IMEX) are often a good choice for hyperbolic systems with stiff source. We shall describe IMEX RK-finite-difference schemes obtained by a method of lines approach.

1.1 Hyperbolic Systems

Let us consider a system of equations of the form

$$\frac{\partial u}{\partial t} + \frac{\partial f(u)}{\partial x} = 0, \tag{1.1}$$

where $u(x,t) \in \mathbb{R}^{m_c}$ is the unknown vector field, and $f : \mathbb{R}^{m_c} \to \mathbb{R}^{m_c}$ is assumed to be a smooth function. The system is hyperbolic in the sense that for any $u \in \mathbb{R}^{m_c}$, the Jacobian matrix $A = \nabla_u f(u)$ has real eigenvalues and its eigenvectors span $\mathbb{R}^{m_c}$.

Such a system is linear if the Jacobian matrix does not depend on u, otherwise it is called quasilinear.

Linear hyperbolic systems are much easier to study. For these systems, the initial value problem is well posed, and the solution maintains the regularity of the initial data for any time. Such systems can be diagonalized, and therefore they can be reduced to m_c linear scalar equations.

The situation is much different with quasilinear systems. For them the initial value problem is well posed locally in time. In general, the solution loses the regularity of the initial data after finite time. Even in the case of the single scalar equation, i.e., $m_c = 1$, the strong solution ceases to exist, and it is necessary to consider weak solutions. For smooth initial data, these have the general appearance of piecewise smooth functions, which contain jump discontinuities [36].

If we denote by x_Σ the position of the discontinuity and by V_Σ its velocity, then the jump conditions across Σ read

$$-V_\Sigma [\![u]\!] + [\![f]\!] = 0, \tag{1.2}$$

where for any quantity $h(x)$, $[\![h]\!] = h(x_\Sigma^+) - h(x_\Sigma^-)$ denotes the jump across the discontinuity interface Σ.

As an example, Figure 1.1 shows the solution of Burgers equation, for which $m_c = 1$ and $f = u^2/2$, with the initial condition

$$u(x,0) = 1 + \frac{1}{2}\sin(\pi x), \tag{1.3}$$

$x \in [-1,1]$, and periodic boundary conditions. A discontinuity forms at time $t = 2/\pi \approx 0.6366$. The figure shows the initial condition, and the solution at times $t = 0.5$ and $t = 0.8$. In the latter case, the parametric solution constructed by the characteristics is multi-valued. A single-valued solution is restored by fitting a shock discontinuity at a position that maintains conservation. An excellent introduction to the subject of linear and nonlinear waves is the classical book by Whitham [64].

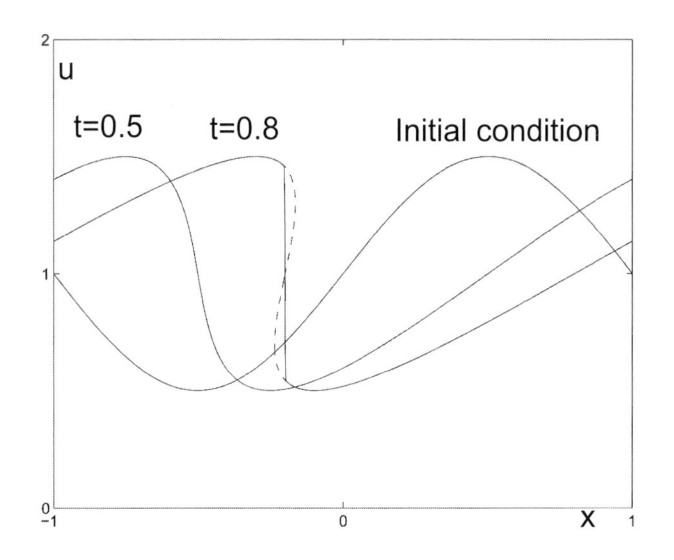

Figure 1.1: Burgers equation at different times ($t = 0, 0.5, 0.8$).

Piecewise smooth solutions that satisfy the jump conditions are not unique (see, for example, [38], Section 3.5).

An entropy condition is used to guarantee uniqueness of the solution, at least in the scalar case. It states that for any convex function $\eta(u)$ there exists an entropy flux $\psi(u)$ such that the pair $[\eta, \psi]$ satisfies (see Section 1.6)

$$\frac{\partial \eta}{\partial t} + \frac{\partial \psi(u)}{\partial x} \leq 0 \tag{1.4}$$

for any weak solution of the equation, and the equal sign holds for smooth solutions.

In the scalar case the entropy condition ensures that the weak solution is the unique viscosity solution, i.e., it is obtained as the limit

$$\lim_{\epsilon \to 0} u^\epsilon(x, t),$$

where u^ϵ satisfies the equation

$$\frac{\partial u^\epsilon}{\partial t} + \frac{\partial f(u^\epsilon)}{\partial x} = \epsilon \frac{\partial^2 u^\epsilon}{\partial x^2}.$$

For the relation between entropy condition and viscosity solutions in the case of systems see, for example, [36], or [16].

The mathematical theory of hyperbolic systems of conservation laws is used as a guideline for the construction of schemes for the numerical approximation of conservation laws. Consider, for example, the conservation property. Integrating Equation (1.1) over an interval $[a, b]$ one has

$$\frac{d}{dt}\int_a^b u(x,t)\,dx = f(u(a,t)) - f(u(b,t)).$$

If $u(a,t) = u(b,t)$ (for example if the boundary conditions are periodic), then the quantity $\int_a^b u(x,t)$ is conserved in time. Such a conservation property is directly related to the jump condition (1.2).

It is important that a similar conservation property is maintained at a discrete level in a shock-capturing scheme, otherwise the scheme will not provide the correct propagation speed for the discontinuities.

1.2 Simple Three-Point Schemes

In this section we review some of the basic three-point stencil schemes used for the numerical solution of the scalar advection equation

$$\frac{\partial u}{\partial t} + c\frac{\partial u}{\partial x} = 0. \tag{1.5}$$

Standard centered difference in space and forward Euler in time gives a numerical scheme that converges only under the restriction $\Delta t = O(\Delta x^2)$, which is not natural for hyperbolic problems. One can show that using three-level Runge-Kutta time discretization coupled with centered difference leads to a scheme which is stable for $c\Delta t < K\Delta x$, where K is a constant that depends on the scheme. On the other hand, it is possible to combine space and time discretization, and obtain one-level schemes (in time) which are stable. The *upwind scheme* uses a first-order approximation of the space derivative. Upwinding time discretization is obtained by discretizing the space derivatives as follows:

$$\left.\frac{\partial u}{\partial x}\right|_{x_j} \approx \begin{cases} \dfrac{u_j - u_{j-1}}{\Delta x} & \text{if } c \geq 0, \\ \dfrac{u_{j+1} - u_j}{\Delta x} & \text{if } c < 0. \end{cases}$$

First-order upwinding is obtained by explicit Euler and first-order upwind space discretization[1] :

$$\frac{u_j^{n+1} - u_j^n}{\Delta t} + c\frac{u_j^n - u_{j-1}^n}{\Delta x} = 0 \quad \text{if } c \geq 0,$$

[1]For the moment we do not specify the range of the cell index j, which depends also on the boundary conditions and on the order of accuracy of the method.

$$\frac{u_j^{n+1} - u_j^n}{\Delta t} + c\frac{u_{j+1}^n - u_j^n}{\Delta x} = 0 \quad \text{if } c < 0.$$

The scheme can be written in a compact form as

$$u_j^{n+1} = u_j^n - \mu(c_+(u_j^n - u_{j-1}^n) + c_-(u_{j+1}^n - u_j^n)), \tag{1.6}$$

where $\mu = \Delta t/\Delta x$ is the mesh ratio, and for all $x \in \mathbb{R}$, $x_+ \equiv \max(x, 0)$, $x_- \equiv \min(x, 0)$. Let us assume for the sake of argument that $c > 0$, and let us study the consistency of the upwind scheme, by applying the discrete operator to the exact solution of the equation

$$\mathfrak{L}_\Delta u(x,t) \equiv \frac{u(x, t + \Delta t) - u(x,t)}{\Delta t} + c\frac{u(x,t) - u(x - \Delta x, t)}{\Delta x}. \tag{1.7}$$

Let $k \equiv \Delta t$, $h \equiv \Delta x$. By using Taylor expansion of $u(x,t)$ in t and x, one has

$$\frac{u(x, t+k) - u(x,t)}{k} = \frac{\partial u}{\partial t}(x,t) + \frac{k}{2}\frac{\partial^2 u}{\partial t^2}(x, t+\tau),$$

$$\frac{u(x,t) - u(x-h,t)}{h} = \frac{\partial u}{\partial x}(x,t) - \frac{h}{2}\frac{\partial^2 u}{\partial x^2}(x - \xi, t),$$

with $\tau \in [0, k]$, $\xi \in [0, h]$. Inserting this expansion into the discrete operator one has

$$\mathfrak{L}_\Delta u(x,t) = \frac{\partial u}{\partial t} + c\frac{\partial u}{\partial x} + \frac{k}{2}\frac{\partial^2 u}{\partial t^2}(x, t+\tau) - \frac{ch}{2}\frac{\partial^2 u}{\partial x^2}(x - \xi, t).$$

If u satisfies the equation, then $u_t + cu_x = 0$, and the consistency error is given by

$$d(x,t) = \frac{k}{2}\frac{\partial^2 u}{\partial t^2}(x, t+\tau) - \frac{ch}{2}\frac{\partial^2 u}{\partial x^2}(x - \xi, t) = O(k, h).$$

A scheme is said to be *consistent* if $d(x,t) \to 0$ as $k \to 0$, $h \to 0$. In this case the scheme is consistent to the first order in k and h, since $d(x,t)$ is an infinitesimal of first order in Δt and Δx.

The stability of the scheme can be checked in various ways. Since the equation and the method are linear, Fourier analysis can be used. Here we assume periodic boundary conditions. We look for a solution of the form

$$u_j^n = \rho^n e^{ij\xi}, \tag{1.8}$$

where $i = \sqrt{-1}$ denotes the imaginary unit. Assume $c > 0$. Then we have

$$u_j^{n+1} = u_j^n - \frac{k}{h}c(u_j^n - u_{j-1}^n),$$

substituting (1.8) in the above expression one computes the amplification factor

$$\begin{aligned}\rho &= 1 - \frac{k}{h}c(1 - e^{i\xi}) \\ &= 1 - \lambda(1 - \cos\xi) - i\lambda\sin\xi,\end{aligned}$$

where $\lambda \equiv kc/h$. Performing the calculations one has

$$|\rho|^2 = 1 - 2\lambda(1-\lambda)(1-\cos\xi).$$

Because $1-\cos\xi \geq 0$ and $\lambda > 0$, it is $|\rho|^2 < 1$ if $1-\lambda > 0$, and therefore if $\lambda < 1$.

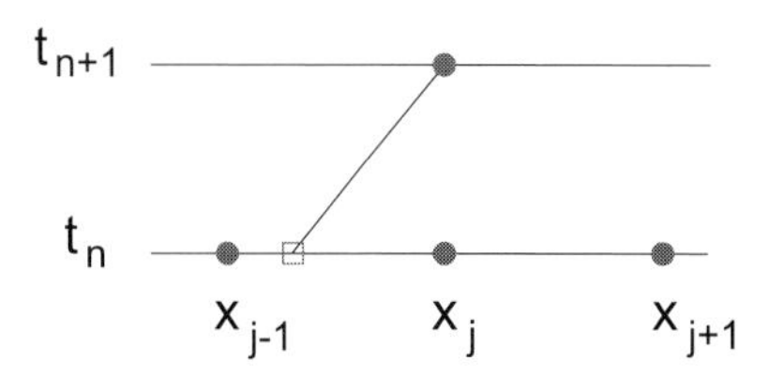

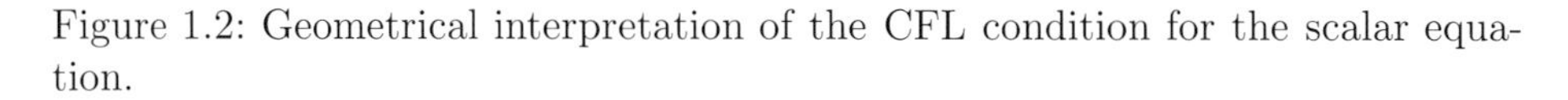

Figure 1.2: Geometrical interpretation of the CFL condition for the scalar equation.

The geometric interpretation of this condition

$$0 < \lambda < 1, \quad \text{i.e.,} \quad 0 < \frac{ck}{h} < 1,$$

is expressed as follows: the characteristics that passes through point (x_j, t_{n+1}), when drawn backward in time, intercepts the line $t = t_n$ at a point between x_{j-1} and x_j (see Figure 1.2). As we shall see, this stability condition is a particular case of a more general necessary condition known as CFL condition.

Another simple three-point scheme is *Lax-Friedrichs scheme* (LxF)

$$u_j^{n+1} = \frac{1}{2}(u_{j+1}^n + u_{j-1}^n) - \frac{ck}{2h}(u_{j+1}^n - u_{j-1}^n), \tag{1.9}$$

which is similar to the explicit Euler + central difference, except that $(u_{j+1}^n + u_{j-1}^n)/2$ replaces u_j^n. Let us check the consistency of the scheme. The discrete operator corresponding to LxF is

$$\begin{aligned} \mathfrak{L}_\Delta u(x,t) = & \frac{1}{k}\left[u(x,t+k) - \frac{1}{2}(u(x-h,t) + u(x+h,t))\right] \\ & + \frac{c}{2h}(u(x+h,t) - u(x-h,t)). \end{aligned}$$

It is easy to check that

$$\mathfrak{L}_\Delta u(x,t) = u_t + cu_x + \frac{1}{2}k\left(u_{tt} - \frac{h^2}{k^2}u_{xx}\right) + O(h^2, k^2). \tag{1.10}$$

Therefore if we assume that $h \to 0$, $k \to 0$, $k/h = \mu =$ constant, then, if u is a solution of the wave equation $u_t + cu_x = 0$, it follows that

$$\mathfrak{L}_\Delta u = \frac{1}{2}k\left(u_{tt} - \frac{h^2}{k^2}u_{xx}\right) + O(h^2, k^2).$$

Therefore the scheme is first-order consistent in h and k. Let us compare upwind and Lax-Friedrichs scheme. Using the property

$$u_{tt} = c^2 u_{xx}\,,$$

one has

- upwind $\qquad \mathcal{L}_\Delta u = \dfrac{1}{2}(kc^2 - ch)u_{xx} = \dfrac{1}{2}ch\left(\dfrac{ck}{h} - 1\right)u_{xx} + O(h^2),$

- LxF $\qquad \mathcal{L}_\Delta u = \dfrac{1}{2}(kc^2 - \dfrac{h^2}{k})u_{xx} = \dfrac{1}{2}\dfrac{h}{k}h\left(\dfrac{c^2k^2}{h^2} - 1\right)u_{xx} + O(h^2).$

Let $\lambda \equiv ch/k$. Then, neglecting second-order terms, we can write

$$\mathcal{L}_{\Delta up}u = \frac{1}{2}ch(\lambda - 1)u_{xx},$$

$$\mathcal{L}_{\Delta LxF}u = \frac{1}{2}ch\frac{\lambda^2 - 1}{\lambda}u_{xx} = \frac{1+\lambda}{\lambda}\mathcal{L}_{\Delta up}u\,.$$

From this expression we see that upwind scheme has in general a smaller discretization error. Furthermore, if $\lambda \to 0$, $\mathcal{L}_{\Delta up} = O(h)$ uniformly, while for fixed h, $\mathcal{L}_{\Delta LxF}$ may be unbounded. Therefore, for a fixed grid size in x, it is not possible to choose arbitrarily small time steps for the LxF scheme.

Stability of LxF scheme can be checked by the usual Fourier technique. One has (see (1.9))

$$\begin{aligned}\rho &= \frac{1}{2}(e^{i\xi} + e^{-i\xi}) - \lambda\frac{e^{i\xi} - e^{-i\xi}}{2}\\ &= \cos\xi - \lambda i \sin\xi\,.\end{aligned}$$

Therefore

$$|\rho|^2 = \cos^2\xi + \lambda^2\sin^2\xi = 1 + (\lambda^2 - 1)\sin^2\xi\,.$$

It is

$$\max_\xi |\rho| = \max(1, |\lambda|)\,,$$

and therefore if $\lambda \leq 1$, no Fourier mode is amplified, and if $|\lambda| > 1$, there are Fourier modes that grow in time. We conclude that a necessary and sufficient condition for the linear stability of LxF scheme is $|\lambda| \leq 1$.

Remark 1.1. Upwind and Lax-Friedrichs schemes are both first-order accurate in space and time, and have the same stability restriction, but upwind has a lower discretization error, therefore it is more accurate. This larger accuracy, however, is counterbalanced by the greater complexity of the scheme. In the case of the single scalar equation, upwinding requires to check for the sign of c. As we shall see, for systems upwinding-based schemes are more complex to apply then central-based schemes, such as LxF.

Second-order accuracy in space and time: the Lax-Wendroff scheme. A simple second-order scheme can be obtained by expanding the solution in time using a Taylor expansion:

$$u(x,t+k) = u + ku_t + \frac{1}{2}k^2 u_{tt} + O(k^3),$$

where we omit the argument of the functions if it is (x,t). Making use of the equation $u_t = -cu_x$, one has

$$u(x,t+k) = u - kcu_x + \frac{1}{2}k^2c^2u_{xx} + O(k^3).$$

Discretizing the space derivatives by central differences and neglecting terms of $O(k^3 + kh^2)$ one has

$$u_j^{n+1} = u_j^n - \frac{\lambda}{2}(u_{j+1}^n - u_{j-1}^n) + \frac{1}{2}\lambda^2(u_{j+1}^n - 2u_j^n + u_{j-1}^n). \tag{1.11}$$

It is easy to check that the consistency error is

$$\mathfrak{L}_{\Delta LW}u(x,t) = u_t + cu_x + \frac{1}{2}k(u_{tt} - c^2u_{xx}) + \frac{ch^2}{6}u_{xxx} + \frac{k^2}{6}u_{ttt} + O(h^2k),$$

therefore, if $u_t + cu_x = 0$, then

$$\mathfrak{L}_{\Delta LW}u(x,t) = \frac{ch^2}{6}u_{xxx} + \frac{k^2}{6}u_{ttt} + O(h^2k),$$

showing that the scheme is actually second-order accurate.

About the stability, Fourier analysis gives

$$\begin{aligned}\rho &= 1 - \lambda\frac{e^{i\xi} - e^{-i\xi}}{2} + \lambda^2(\cos\xi - 1)\\ &= 1 - i\lambda\sin\xi + \lambda^2(\cos\xi - 1)\\ |\rho|^2 &= 1 - \lambda^2(1-\lambda^2)(1-\cos\xi)^2\,,,\end{aligned}$$

and therefore

$$|\rho|^2 \le 1 \quad \forall\xi \in [0,2\pi] \Leftrightarrow |\lambda| \le 1.$$

In Section 3.2 we shall make use of a different concept of stability, more suited for nonlinear equations.

1.3 Conservation Form, Jump Conditions and Conservative Schemes

Let us consider the system (1.1) that we rewrite here

$$u_t + f(u)_x = 0. \tag{1.12}$$

If $f(u)$ is nonlinear, then discontinuities may develop in finite time. For a smooth solution, the system is equivalent to

$$u_t + A(u)u_x = 0, \tag{1.13}$$

where $A(u) = df/du$ is the Jacobian matrix. However, when discontinuities develop, the two forms are no longer equivalent.

A jump discontinuity, once formed, moves with a speed V_Σ given by the jump conditions (1.2).

If we want that such a conservation property persists also at a discrete level, yielding the correct propagation speed of discontinuities, then the numerical scheme has to take a special form (conservative scheme). We distinguish between semidiscrete and fully discrete conservative schemes. They can be interpreted as an approximation of the conservation equation in integral form.

The general approach to construct semidiscrete schemes is obtained as follows. Let us divide space into J cells $I_j \equiv [x_{j-1/2}, x_{j+1/2}]$ centered at x_j, $j = 1, \dots, J$ (see Figure 1.3). Let us integrate Equation (1.12) in cell I_j, and divide by Δx. Then we obtain:

$$\frac{d\bar{u}_j}{dt} + \frac{1}{\Delta x}[f(u(x_{j+1/2}, t)) - f(u(x_{j-1/2}, t))] = 0, \tag{1.14}$$

where

$$\bar{u}_j(t) \equiv \frac{1}{\Delta x}\int_{x_{j-1/2}}^{x_{j+1/2}} u(x,t)\,dx$$

denotes the cell average.

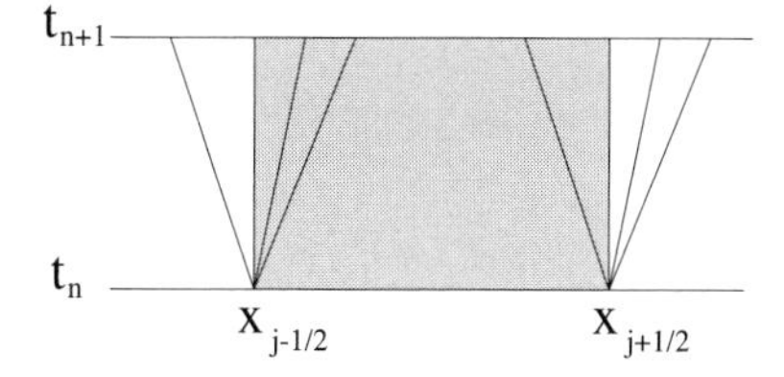

Figure 1.3: Integration over a cell in space-time and definition of finite-volume, fully discrete schemes.

Equation (1.14) suggests to look for numerical schemes of the form

$$\frac{d\bar{u}_j}{dt} = -\frac{1}{\Delta x}[F_{j+1/2} - F_{j-1/2}], \tag{1.15}$$

where $F_{j+1/2}$ is the numerical flux at the right edge of the cell j.

In order to convert (1.14) into a numerical scheme, one needs to relate pointwise values of $u(x,t)$ with cell averages. In this way $F_{j+1/2}$ will be some function of $\bar{u}_j$ and $\bar{u}_{j+1}$ (and possibly of other cell averages, in more complex schemes). In the simplest case one has

$$F_{j+1/2} = F(\bar{u}_j, \bar{u}_{j+1}).$$

Scheme (1.15), with some choice of the *numerical flux function* $F(\cdot,\cdot)$, is in semidiscrete form. Time discretization can be used to solve the system of ODE's (1.15). This approach allows a great flexibility in the choice of the numerical flux function and time integration. This approach of performing first space and then time discretization is called *method of lines*. It is the probably the preferred choice for developing very high-order, general purpose schemes.

Fully discrete schemes are obtained either by discretizing the semidiscrete scheme in time (method of lines) or directly by integrating Equation (1.12) in a cell in space and time (see Figure 1.3). Integration over the cell gives

$$\bar{u}_j^{n+1} = \bar{u}_j^n - \frac{1}{\Delta x}\int_{t_n}^{t_{n+1}} [f(u(x_{j+1/2},t)) - f(u(x_{j-1/2},t))]\,dt.$$

A numerical scheme inspired by this form of the equation takes the form

$$\bar{u}_j^{n+1} = \bar{u}_j^n - \frac{\Delta t}{\Delta x}(F_{j+1/2}^n - F_{j-1/2}^n), \tag{1.16}$$

where now the numerical flux $F_{j+1/2}^n$ is an approximation of the time average of f along the edge of the cell average in the two adjacent cells. In the simplest case one has

$$F_{j+1/2}^n = F(\bar{u}_j^n, \bar{u}_{j+1}^n). \tag{1.17}$$

The schemes we have seen (upwind, LxF and LW) belong to this category. For example, the upwind scheme applied to the scalar case of equation (1.12), assuming $c(u) = df/du > 0$, takes the form

$$\bar{u}_j^{n+1} = \bar{u}_j^n - \frac{\Delta t}{\Delta x}(f(\bar{u}_j^n) - f(\bar{u}_{j-1}^n)),$$

therefore $F_{j+1/2}^{up} = f(\bar{u}_j^n)$.

The Lax-Friedrichs scheme takes the form

$$\bar{u}_j^{n+1} = \frac{1}{2}(\bar{u}_{j+1}^n + \bar{u}_{j-1}^n) - \frac{\Delta t}{2\Delta x}(f(\bar{u}_{j+1}^n) - f(\bar{u}_{j-1}^n)).$$

This can be written as

$$\bar{u}_j^{n+1} = \bar{u}_j^n - \frac{\Delta t}{\Delta x}[F_{j+1/2}^n - F_{j-1/2}^n],$$

with

$$F_{j+1/2}^n = F_{j+1/2}^{LxF} = \frac{1}{2}(f(\bar{u}_{j+1}^n) + f(\bar{u}_j^n)) - \frac{\Delta x}{2\Delta t}(\bar{u}_{j+1}^n - \bar{u}_j^n). \tag{1.18}$$

The Lax-Wendroff scheme can also be written in this form. In fact, the Lax-Wendroff scheme is based on Taylor expansion, therefore

$$\begin{aligned} u(x,t+k) &= t + ku_t + \frac{1}{2}k^2 u_{tt} + O(k^3) \\ &= u - kf(u)_x - \frac{1}{2}k^2 f(u)_{tx} + O(k^3) \\ &= u - k\left[f(u) + \frac{1}{2}k(Au_t)\right]_x + O(k^3) \\ &= u - k\left[f(u) - \frac{1}{2}kAf(u)_x\right]_x + O(k^3). \end{aligned}$$

Neglecting $O(k^3)$ terms, this expression suggests the following conservative scheme:

$$\bar{u}_j^{n+1} = \bar{u}_j^n - \frac{\Delta t}{\Delta x}\left[F_{j+1/2}^{LW} - F_{j-1/2}^{LW}\right],$$

with

$$F_{j+1/2}^{LW} = \frac{1}{2}(f(\bar{u}_j^n) + f(\bar{u}_{j+1}^n)) - \frac{\Delta t}{2\Delta x}A_{j+1/2}(f(\bar{u}_{j+1}) - f(\bar{u}_j)),$$

where the Jacobian matrix $A_{j+1/2}$ is the matrix $A(u)$ computed in $(\bar{u}_j + \bar{u}_{j+1})/2$.

Remark 1.2. The Lax-Wendroff scheme requires the calculation of the Jacobian matrix. Second-order explicit schemes that do not require the calculation of A are the Richtmyer two-step Lax-Wendroff method (RLW)

$$\begin{aligned} u_{j+1/2}^{n+1/2} &= \frac{1}{2}(u_j^n + u_{j+1}^n) - \frac{\Delta t}{\Delta x}(f(u_{j+1}^n) - f(u_j^n)), \\ u_j^{n+1} &= u_j^n - \frac{\Delta t}{\Delta x}(f(u_{j+1/2}^{n+1/2}) - f(u_{j-1/2}^{n+1/2})), \end{aligned}$$

and the MacCormack, method (MC)

$$\begin{aligned} u_j^* &= u_j^n - \frac{\Delta t}{\Delta x}(f(u_{j+1}^n) - f(u_j^n)), \\ u_j^{n+1} &= \frac{1}{2}(u_j^n + u_j^*) - \frac{\Delta t}{2\Delta x}(f(u_j^*) - f(u_{j-1}^*)). \end{aligned}$$

Note that all these methods (LW, RLW, MC) are second-order accurate in space and time, and the amplification factor is the same, namely

$$\rho = 1 - \lambda^2(1 - \cos\xi) - i\lambda\sin\xi,$$

therefore they are linearly stable if $|\lambda_{\max}| \le 1$, where

$$\lambda_{\max} = \max_j c(u_j^n)\frac{\Delta t}{\Delta x}$$

1.4 Consistency and Convergence

For a conservative scheme to be consistent, one requires that

$$\mathcal{L}_\Delta u \to 0 \quad \text{as} \;\; h,k \to 0,\; h/k \;\; \text{fixed}.$$

The condition

$$F(v,v) = f(v) \qquad \forall v \in \mathbb{R}^{m_c}, \tag{1.19}$$

together with some regularity assumption on F (for example that F is Lipschitz continuous in both arguments) ensures consistency of the scheme (1.16)–(1.17). In fact one has

$$\mathcal{L}_\Delta u(x,t) = \frac{u(x,t+k)-u(x,t)}{k} + \frac{F(u(x,t),u(x+h,t)) - F(u(x-h,t),u(x,t))}{h}.$$

Using Taylor expansion, and assuming differentiability of F, one has

$$\mathcal{L}_\Delta u(x,t) = u_t + \left(\frac{\partial F}{\partial u_1} + \frac{\partial F}{\partial u_2}\right) u_x + O(h,k),$$

where by $\partial F/\partial u_1$, $\partial F/\partial u_2$ we denote the derivatives with respect to the first and second argument, respectively, and the functions are computed at $u(x,t)$. Condition (1.19) and differentiability imply

$$\frac{\partial F}{\partial u_1} + \frac{\partial F}{\partial u_2} = \frac{\partial f}{\partial u},$$

and therefore $\mathcal{L}_\Delta u(x,t) = O(h,k)$ for smooth solutions of the conservation law (1.12).

A more useful concept of stability, that can be applied also to the nonlinear case, is the L^1-stability. We shall discuss it in Section 3.2.

1.5 Conservation Properties

Let us assume periodic boundary conditions at the edge of the interval $[a,b]$. Integrating the conservation law over a period, one has

$$\frac{d}{dt}\int_a^b u(x,t)dx = 0,$$

since $u(a,t) = u(b,t)$.

We assume that the interval is divided into J cells of size $\Delta x = (b-a)/J$.

If a conservative scheme is used, then such a property is maintained at a discrete level. In fact, from

$$u_j^{n+1} = u_j^n - \frac{\Delta t}{\Delta x}\left[F_{j+1/2} - F_{j-1/2}\right],$$

one has, summing over all the J cells,

$$\sum_{j=1}^{J} u_j^{n+1} = \sum_{j=1}^{J} u_j^n - \frac{\Delta t}{\Delta x}\left[F_{J+1/2} - F_{J-1/2} + F_{J-1/2} - F_{J-3/2} + \cdots + F_{3/2} - F_{1/2}\right]$$

$$= \sum_{j=1}^{J} u_j^n,$$

since all the intermediate terms cancel, and $F_{J+1/2} = F_{1/2}$ because of periodicity ($F_{J+1/2} = F(u_J, u_{J+1}) = F(u_0, u_1) = F_{1/2}$, since $u_0 = u_J$, and $u_{J+1} = u_1$).

Consistency also suggests that a conservative scheme gives the correct speed of propagation for discontinuities. A more mathematical justification for the use of conservative schemes is given by the Lax-Wendroff theorem, which states that if a discrete solution converges to a function $u(x,t)$, then this function is a weak solution of the conservation law [37].

1.6 Entropy Condition

It is well known that weak solutions to conservation laws are not unique, even in the case of the scalar equation.

Uniqueness, however, can be restored if an additional condition is imposed. This additional condition is called *entropy condition.*

What is a mathematical entropy? An entropy function can be defined both for the scalar equation and for systems of conservation laws. It is a convex function of the unknown field u that satisfies an additional conservation law for smooth solutions. A convex function $\eta(u)$ is an entropy associated to Equation (1.12) if there exists a function $\psi(u)$, called *entropy flux*, such that, for all smooth solutions of the conservation equation (1.12), the following equation is satisfied:

$$\eta(u)_t + \psi(u)_x = 0. \tag{1.20}$$

When does a system admit an entropy? For smooth solutions the above equation can be written as

$$\eta'(u)u_t + \psi'(u)u_x = 0.$$

Compatibility of this equation with the balance equation (1.12) for u,

$$u_t + f'(u)u_x = 0,$$

requires that

$$\psi'(u) = \eta'(u)f'(u). \tag{1.21}$$

This is in fact the condition that is used to construct the entropy flux. Now, for a scalar equation it is easy to satisfy this condition since given an arbitrary convex function $\eta(u)$, then $\psi(u)$ is a primitive of $\eta'(u)f'(u)$. For a 2×2 system, (1.21) constitutes a set of two equations (one for each component of u) in two unknown functions. Therefore it is conceivable to expect that, in general, such a system admits solutions. In this respect, 2×2 system are very special, and in fact there are general theorems that hold for a generic hyperbolic 2×2 system, but not for a generic hyperbolic $m \times m$ system. In the latter case, in fact, one has a set of m_c partial differential equations for only two unknown functions, η and ψ, and therefore in general the existence of an entropy function is not guaranteed.

We remark that most systems of conservation laws that come from physics admit an entropy function. Gas dynamics, for example, is a system of three equations and admits a mathematical entropy that is related to the physical entropy of the gas.

What does the existence of an entropy function guarantee? Existence of a convex entropy is a very important property of a hyperbolic system of conservation laws, since if a system possesses a convex entropy then it is symmetrizable, i.e., there exists an invertible transformation of field variables such that in the new variables the system is symmetric. An account of the theory can be found in the classical paper by Friedrichs and Lax [19] and in the book by Courant and Friedrichs [15].

Vanishing Viscosity

One method to restore the uniqueness of the solution for a scalar equation or for a system of conservation laws is to consider the weak solution of the system which is a limit of a sequence of solutions to the regularized equation, as the regularizing parameter vanishes. Consider the equation

$$\begin{aligned} &\frac{\partial u^\varepsilon}{\partial t} + \frac{\partial f(u^\varepsilon)}{\partial x} = \varepsilon \frac{\partial^2 u}{\partial x^2} \quad && x \in \Omega, \ \ t \in [0, T], \\ &u(x, 0) = u_0(x) && x \in \Omega. \end{aligned} \tag{1.22}$$

For any $\varepsilon > 0$, the initial value problem Equation (1.22) has a regular unique solution $u^\varepsilon(x, t)$. It is possible to show that $u = \lim_{\varepsilon \to 0} u^\varepsilon$ is a weak solution of the original equation (1.12). Such a solution, whose uniqueness is a consequence of the properties of problem (1.22), is called the *viscosity solution* of Equation (1.12). It is possible to show that across a discontinuity, the entropy-entropy flux satisfies the inequality

$$-V \llbracket \eta \rrbracket + \llbracket \psi \rrbracket \leq 0, \tag{1.23}$$

while in the smooth regions equation (1.20) is satisfied. Both conditions can be summarized by one inequality, called *entropy inequality*:

$$\eta(u)_t + \psi(u)_x \leq 0. \tag{1.24}$$

The above expression has to be interpreted in the weak sense, i.e.,

$$\int_0^T \int_a^b (\eta(u)\phi_t + \psi(u)\phi_x)dx\,dt + \int_a^b \eta(u_0(x))\phi(x,0)dx \geq 0, \tag{1.25}$$

where $\phi(x,t)$ is an arbitrary regular test function that vanishes at $x = a$ and $x = b$ and at $t = T$. A weak solution satisfying the entropy inequality (1.24) is called an *entropy solution* of the conservation law (1.1).

From what we said above, if a function u is a viscosity solution then it is an entropy solution. Is the converse true? This depends on the uniqueness of the entropy solution. It can be proved that uniqueness holds in the case of the scalar equation, while for arbitrary systems the question is still open. For most systems of physical relevance that admit an entropy, uniqueness of the entropy solution can be proved.

A more detailed treatment of this subject can be found, for example, in the book by Godlewski and Raviart [21] or in the book by Dafermos [16].

1.7 Discrete Entropy Condition

Some numerical schemes possesses a discrete entropy-entropy flux pair which helps to select the correct entropy solution, as the grid is refined.

Discrete entropy conditions can be written as

$$\eta(u_j^{n+1}) \leq \eta(u_j^n) - \frac{\Delta t}{\Delta x}\left[\Psi(u_j^n, u_{j+1}^n) - \Psi(u_{j-1}^n, u_j^n)\right], \tag{1.26}$$

where $\Psi(u,v)$ is a suitable numerical entropy flux, consistent with the flux function ψ, i.e.,

$$\Psi(u,u) = \psi(u).$$

We recall that the Lax-Wendroff theorem ensures that if a numerical solution of a conservative scheme applied to a scalar conservation law converges to a function $u(x,t)$, then the function is a weak solution of the equation. If the scheme possesses a discrete entropy inequality, then one is guaranteed that the function $u(x,t)$ is the entropic solution of the equation.

1.8 Dissipation, Dispersion, and the Modified Equation

We have seen that, when the stability condition is satisfied, the amplitude of the Fourier modes decay for Lax-Friedrichs, upwind, and Lax-Wendroff schemes. However, the decay rate for the various schemes is very different. In fact, while $1 - |\rho| = O(h^2)$ for both Lax-Friedrichs and first-order upwind scheme, giving a decay which is locally second order in h (or in Δt) and globally first order, in the case of the Lax-Wendroff scheme one has $1 - |\rho| = O(h^4)$, which gives a decay which is locally fourth order and globally third order in h, in spite of the fact that the scheme is second-order accurate. We may say that, while decay is a primary effect of LxF and upwind schemes (the decay is of the same order of the scheme), it seems to be a very weak effect in the case of the Lax-Wendroff scheme (the decay is of higher order than the scheme).

As we shall see, a different effect will be the dominant one of LW scheme. This behavior, and the qualitative behavior of these numerical schemes can be better understood with the use of the so called *modified equation* associated to the numerical scheme.

Let us consider, for example, the Lax-Friedrichs scheme. The discretization error is given by Equation (1.10). This means that if we apply it to a function u satisfying the equation

$$u_t + cu_x = \frac{1}{2}k\left(\frac{h^2}{k^2}u_{xx} - u_{tt}\right), \tag{1.27}$$

we would have

$$\mathcal{L}_{\Delta LxF}u = O(h^2, k^2).$$

To the same order of accuracy, the function u satisfies the equation

$$u_t + cu_x = \nu_{LF}u_{xx} \tag{1.28}$$

with

$$\nu_{LF} = \frac{1}{2}k\left(\frac{h^2}{k^2} - c^2\right) = \frac{c\,h}{2}\frac{1-\lambda^2}{\lambda}. \tag{1.29}$$

This means that the numerical solution of the Lax-Friedrichs scheme applied to the single equation $u_t + cu_x = 0$ approximates, to a higher degree of accuracy, the solution of Equation (1.28). Such an equation is called the *modified equation* associated to the LxF scheme, and has the structure of a convection-diffusion equation.

The behavior of the scheme applied to the advection equation $u_t + cu_x = 0$ can be qualitatively described by this equation. In particular we observe that

$$\nu_{LF} \geq 0 \quad \Leftrightarrow \quad |\lambda| \leq 1,$$

therefore, the stability condition for the Lax-Friedrichs scheme corresponds to the well-posedness of the initial value problem for the modified equation. For such an equation, a bell-shaped profile becomes broader as it moves, and discontinuities will be smoothed away, as shown in Figure 1.4. Similar things can be said about the upwind scheme (assuming, for example, $df/du > 0$). In this case one has (see section 1.2 on truncation error)

$$\mathcal{L}_{\Delta up}u = u_t + c\,u_x + c\,h(\lambda - 1)u_{xx} + O(h^2, k^2),$$

and therefore the numerical solution obtained by applying the upwind scheme to the scalar equation $u_t + c\,u_x = 0$ approximates, to second-order accuracy, the solution of the equation

$$u_t + c\,u_x = c\,h(1-\lambda)u_{xx},$$

which is again a convection-diffusion equation with diffusion coefficient

$$\nu_{up} = c\,h(1-\lambda).$$

Both schemes (LxF and upwind) are dissipative, the diffusion coefficient (numerical viscosity) is of first order in h (because the scheme is first-order accurate), and vanishes for $\lambda = 1$, since in this case the characteristics pass exactly through the grid points, and the propagation of the signal is exact. This can be obtained only when a scheme is applied to a linear equation, but it has to be kept in mind even when dealing with systems: for fully discrete schemes, best performance (in terms of accuracy and efficiency) is obtained for values of λ close to the maximum value compatible with stability.

The choice of the time step. If the equation is nonlinear, condition $|\lambda| \leq 1$ has to be applied to all cells, and becomes

$$\max_j |c(u_j)| \frac{\Delta t}{\Delta x} \leq 1.$$

This condition, in practice, dictates the choice of the time step. Let $\tilde{\lambda}$ be a constant close to the stability limit (e.g., take $\tilde{\lambda} = 0.95$ for the LxF scheme) then the time step Δt will be chosen as

$$\Delta t = \tilde{\lambda} \frac{\Delta x}{\max_j |c(u_j)|}. \tag{1.30}$$

Using the maximum time step compatible with stability will improve accuracy, since the $O(h)$ term of the discretization error will be as small as possible, and efficiency, since fewer steps will be required to reach the final time T.

For problems for which $\max_j |c(u_j)|$ changes a lot with time, it is extremely important to use an optimal time step according to a relation such as (1.30). The parameter $\tilde{\lambda}$ appearing in (1.30) is often called *Courant number*.

We remark that $\nu_{up} < \nu_{LF}$, and that $\nu_{LF} \to \infty$ as $\lambda \to 0$, which means that for a fixed space grid the discretization error is unbounded for arbitrarily small values of the time step.

1.9 Second-Order Methods and Dispersion

Let us consider the Lax-Wendroff scheme. The consistency error is given by

$$\mathcal{L}_\Delta u = L_\Delta u + O(h^2 k),$$

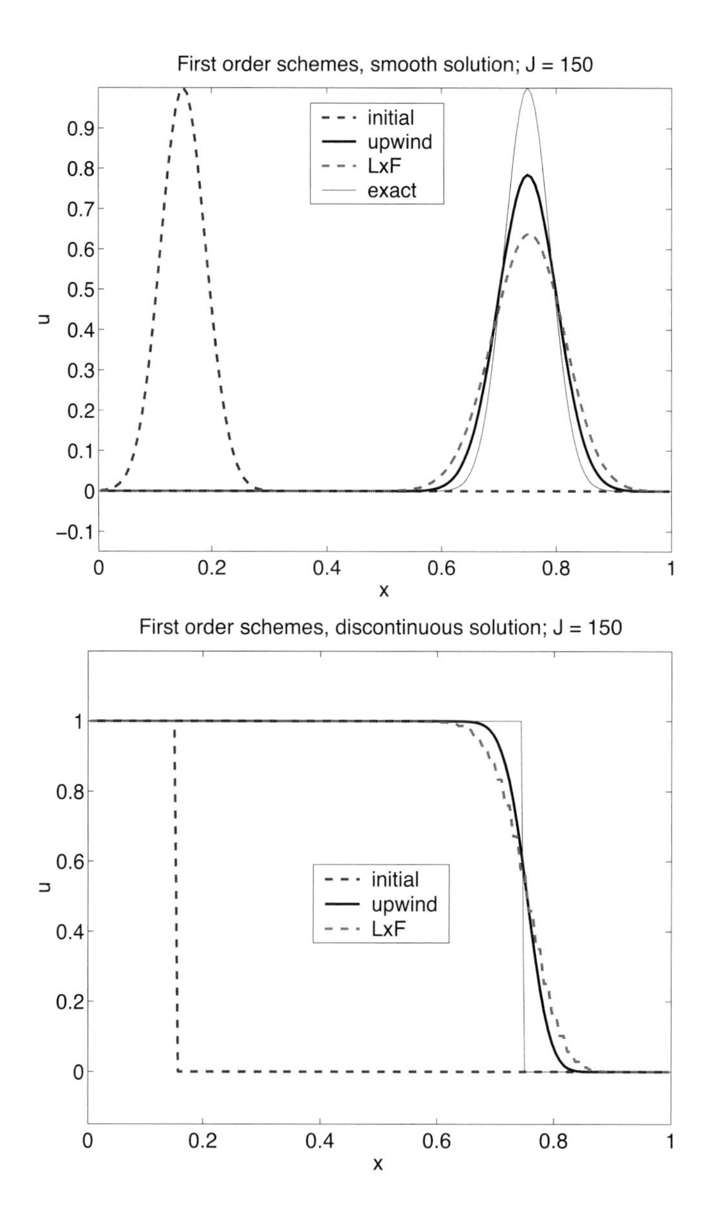

Figure 1.4: Dissipative first-order schemes applied to the linear equation. Top: smooth initial condition. Bottom discontinuous solution. The propagation speed is $c = 1$, and the final time is $T = 0.6$. The number of cells in the calculation is $J = 150$. The solution at the final time represent, in increasing order of dissipation, the exact, the upwind and the LxF solutions. The Courant number $\lambda = 0.75$ has been used.

with

$$L_\Delta u \equiv u_t + c\, u_x + \frac{1}{2}k(u_{tt} - c^2 u_{xx}) + \frac{1}{6}(k^2 u_{ttt} + c\, h^2 u_{xxx}), \tag{1.31}$$

therefore, if we choose u to satisfy

$$L_\Delta u = 0, \tag{1.32}$$

then $\mathcal{L}_\Delta u = O(h^2 k)$. Equation (1.31,1.32) is a modified equation of the Lax-Wendroff scheme. However, its form is not convenient because of the high-order derivative in time. To the same degree of approximation, we can deduce a simpler equation as follows. From Equations (1.31)–(1.32) one has

$$u_t = -c\, u_x - \frac{1}{2}k(u_{tt} - c^2 u_{xx}) + O(k^2).$$

Differentiating in time gives

$$u_{tt} = c^2 u_{xx} - \frac{1}{2}k(u_{ttt} + c^3 u_{xxx}) + O(k^2),$$

$$u_{ttt} = -c^3 u_{xxx} + O(k).$$

From these relations it follows that

$$u_{ttt} + c^3 u_{xxx} = O(k),$$

and therefore

$$u_{tt} - c^2 u_{xx} = O(k^2).$$

Making use of these relations in to (1.31) one has

$$u_t + c u_x - \frac{1}{6}(k^2 c^3 - c\, h^2) u_{xxx} = O(k^3),$$

therefore, neglecting high-order terms, we obtain the modified equation for the Lax-Wendroff scheme,

$$u_t + c\, u_x = \mu_{LW} u_{xxx}, \tag{1.33}$$

with

$$\mu_{LW} \equiv \frac{c\, h^2}{6}(\lambda^2 - 1).$$

This equation is said to have a dispersive character, since small perturbations travel with speed that depends in the frequency. Let us look for an elementary traveling wave solution of the form

$$u(x,t) = \rho \exp(i(\kappa\, x - \omega\, t)),$$

where i is the imaginary unit and κ is the wave number. This is a solution of (1.33) if

$$-i\,\omega + i\,c\,\kappa = -i\mu_{LW}\kappa^3,$$

that is if

$$\omega = c\,\kappa + \mu_{LW}\kappa^3.$$

The ratio ω/κ is called phase velocity, and the derivative $v_g(\kappa) = \partial\omega/\partial\kappa$ is called *group velocity*, and represents the propagation speed of a wave packet centered at wave number κ. The consequence of this dispersive behavior is that an initial profile does not travel unperturbed, since its Fourier components will move at different speeds.

While the LW method is more accurate than first-order schemes, as is shown in the upper part of in Figure 1.5, the dispersive behavior becomes dramatic in the case of an initial discontinuity, and oscillations will appear in the profile, as it is shown in the lower part of the same figure.

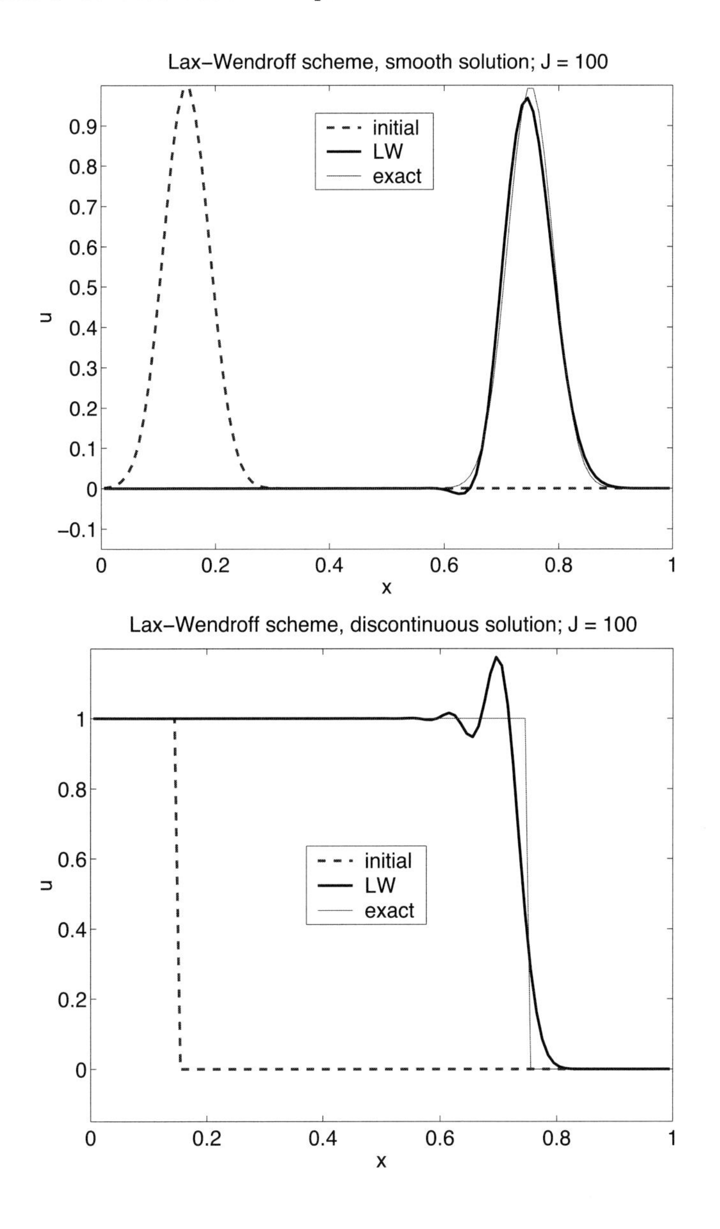

Figure 1.5: Dispersive behavior of the second-order Lax-Wendroff scheme. Top: smooth initial condition. Bottom: discontinuous solution. The propagation speed is $c = 1$, and the final time is $T = 0.6$. $J = 100$ has been used. The CFL number is $c\Delta t/\Delta x = 0.75$. At the final time the exact and the LW solutions are reported.

Chapter 2

Upwind Scheme for Systems

For the scalar conservation law

$$u_t + c\,u_x = 0,$$

first-order upwind is written as

$$u_j^{n+1} = u_j^n - c_+ \frac{\Delta t}{\Delta x}(u_j^n - u_{j-1}^n) - c_- \frac{\Delta t}{\Delta x}(u_{j+1}^n - u_j^n),$$

where

$$c_+ = \max(c, 0), \qquad c_- = \min(c, 0).$$

For a linear system one has:

$$\frac{\partial u}{\partial t} + A\frac{\partial u}{\partial x} = 0, \quad A \in \mathbb{R}^{m_c \times m_c},\ u \in \mathbb{R}^{m_c}, \tag{2.1}$$

and it is not clear where to apply right or left difference.

Upwind schemes for a linear system can be constructed by diagonalizing the system. Let Q be the matrix formed by the m_c independent right eigenvectors of A and Λ the diagonal matrix containing the corresponding eigenvalues. Then one has

$$A\,Q = Q\,\Lambda. \tag{2.2}$$

The diagonalization is always possible if we assume that the system is hyperbolic. Let us express the vector field u as a linear combination of eigenvectors of A:

$$u = Qv, \quad v \in \mathbb{R}^{m_c}. \tag{2.3}$$

Then, substituting into (2.1) one has

$$\frac{\partial v}{\partial t} + \Lambda\frac{\partial v}{\partial x} = 0. \tag{2.4}$$

This means that the equations decouple, and one can apply the upwind scheme to each scalar equation of system (2.4),

$$v_j^{n+1} = v_j^n - \frac{\Delta t}{\Delta x}\left[\Lambda_+(v_j^n - v_{j-1}^n) + \Lambda_-(v_{j+1}^n - v_j^n)\right], \tag{2.5}$$

where

$$\Lambda_+ = \operatorname{diag}\left((\lambda_1)_+, \dots, (\lambda_{m_c})_+\right)$$
$$\Lambda_- = \operatorname{diag}\left((\lambda_1)_-, \dots, (\lambda_{m_c})_-\right),$$

and λ_j, $j = 1, \dots, m_c$ denotes the eigenvalues of A.

Using transformation (2.3) to go back to the original variable u one has the upwind scheme for u in the form

$$u_j^{n+1} = u_j^n - \frac{\Delta t}{\Delta x}\left[A_+(u_j^n - u_{j-1}^n) + A_-(u_{j+1}^n - u_j^n)\right],$$

with

$$A_+ = Q\Lambda_+Q^{-1}, \qquad A_- = Q\Lambda_-Q^{-1}.$$

What is the restriction on the time step? The restriction is that for all eigenvalues λ_ℓ, condition

$$|\lambda_\ell|\frac{\Delta t}{\Delta x} \le 1, \quad \ell = 1, \dots, m_c,$$

has to be satisfied. This condition can be written in the form

$$\rho(A)\frac{\Delta t}{\Delta x} \le 1,$$

where

$$\rho(A) \equiv \max_{1\le j\le m_c} |\lambda_j(A)|$$

denotes the spectral radius of matrix A, i.e., the maximal eigenvalue of the matrix (in absolute value).

The geometric interpretation of the stability condition is the following. For each eigenvalue λ_l one uses left or right difference on the characteristic variable according to whether $\lambda_l > 0$ or $\lambda_l < 0$. The characteristic emanating back from point (x_j, t_{n+1}) will intercept the line $t = t_n$ at a point which lies between x_{j-1} and x_j (if $\lambda_l > 0$) or between x_j and x_{j+1} (if $\lambda_l < 0$) (see Figure 2.1).

This condition is a particular case of a more general stability condition for systems, known as CFL condition, which states that a necessary condition for stability is that the analytical domain of dependence of a given grid point has to be contained by the numerical domain of dependence. Using the same argument of the diagonalization, one can show that a stability condition for the Lax-Friedrichs and the Lax-Wendroff scheme is also

$$\rho(A)\frac{\Delta t}{\Delta x} \le 1.$$

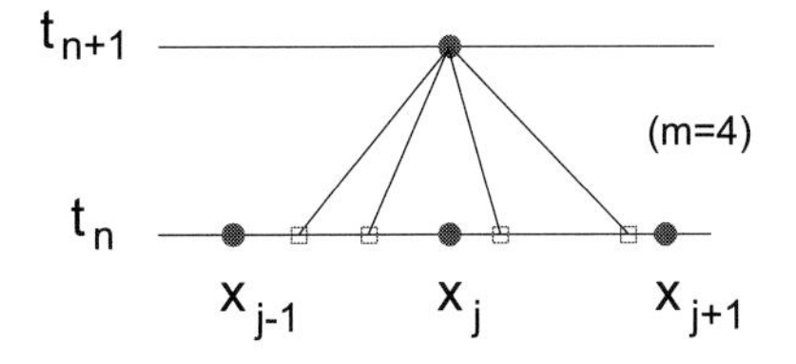

Figure 2.1: Numerical (bullets) and analytical (squares) domain of dependence and geometrical interpretation of the CFL condition.

2.1 The Riemann Problem

How can one generalize upwind schemes to nonlinear conservation laws? A popular method, that can be considered the ancestor of many modern schemes for the numerical approximation of conservative laws, is the Godunov method. This method is based on the solution of the Riemann problem.

A Riemann problem is an initial value problem for which the initial data is piecewise constant:

$$\frac{\partial u}{\partial t} + \frac{\partial f}{\partial x} = 0,$$

$$u(x,0) = \begin{cases} u_l, & x < 0, \\ u_r, & x > 0. \end{cases}$$

For the scalar equation the Riemann problem can be explicitly solved. For example, in the case $f(u) = \frac{1}{2}u^2$ we can have the two cases illustrated in Figure 2.2.

The solution of the Riemann problem is known for several hyperbolic systems of conservation laws with great relevance in the applications, as is the case of gas dynamics (see, for example, [39]). The solution to the Riemann problem centered at the origin is a similarity solution that depends on x/t (see Figure 2.3) $u = u(x/t; u_l, u_r)$. In many cases, however, its solution is not available analytically, or it is quite expensive to compute. In such cases one uses either approximate Riemann solvers, or schemes that do not require the solution to the Riemann problem. For the moment we shall assume that we know the solution of the Riemann problem.

2.2 Godunov Scheme

Let us assume that at time t_n we know an approximation of the cell average, $\{\bar{u}_j^n\}$, and that the solution is a piecewise constant function:

$$u(x, t_n) \simeq \sum_j \bar{u}_j^n \chi_j(x), \tag{2.6}$$

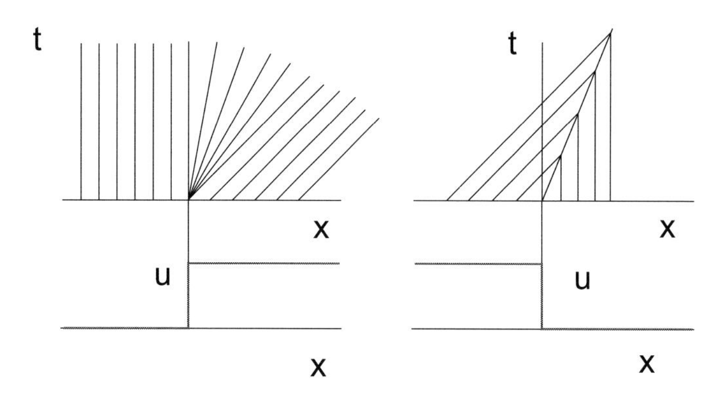

Figure 2.2: Riemann problem: diverging characteristics and rarefaction fan (left), converging characteristics and shock wave (right).

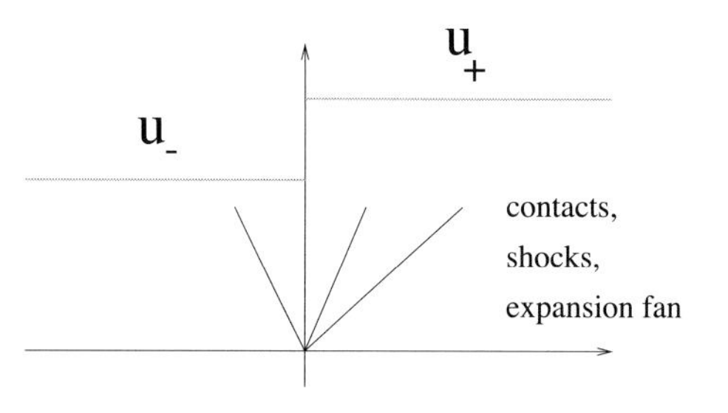

Figure 2.3: Schematic representation of the Riemann fan originating from $(x = 0, t = 0)$.

where

$$\chi_j(x) = \begin{cases} 1 & x \in [x_{j-1/2}, x_{j+1/2}], \\ 0 & \text{otherwise}. \end{cases}$$

For short later times the field vector $u(x,t)$ will be the solution of several Riemann problems, centered in $x_{j+1/2}$.

Let us integrate the conservation law in the cell $I_j \times [t_n, t_{n+1}]$ (see Figure 1.3). Then one has:

$$\bar{u}_j^{n+1} = \bar{u}_j^n - \frac{1}{\Delta x} \int_{t_n}^{t_{n+1}} \left[f(u(x_{j+1/2}, t)) - f(u(x_{j-1/2}, t)) \right] dt. \tag{2.7}$$

Now if the Riemann fan does not interact (which is obtained if the time step Δt satisfies a suitable CFL condition), then the function $u(x_{j+1/2}, t)$ can be obtained from the solution of the Riemann problem with states u_j and u_{j+1} across the interface:

$$u(x_{j+1/2}, t) = u^*(0; \bar{u}_j, \bar{u}_{j+1}) =: u^*(\bar{u}_j, \bar{u}_{j+1}).$$

This quantity does not depend on time, and therefore Equation (2.7) becomes

$$\bar{u}_j^{n+1} = \bar{u}_j^n - \frac{\Delta t}{\Delta x}\left[f(u^*(\bar{u}_j, \bar{u}_{j+1})) - f(u^*(\bar{u}_{j-1}, \bar{u}_j))\right]. \tag{2.8}$$

If the function $u(x, t_n)$ is really a piecewise constant function, then Equation (2.8) gives the correct average of the solution at time t_{n+1}. In order to proceed from time t_{n+1} to time t_{n+2} applying the same technique, one has to approximate the solution at time t_{n+1} as a piecewise constant function. *It is essentially this projection that introduces the approximation.*

When applied to a linear system, the Godunov scheme reduces to first-order upwind. To see this, let us consider an interface, let us say at $x_{j+1/2}$, and let us write

$$[\![\bar{u}]\!]_{j+1/2} = \bar{u}_{j+1} - \bar{u}_j = \sum_k \alpha_k^{(j+1/2)} r_k,$$

where r_k are the right eigenvectors of matrix A that defines the linear flux:

$$f(u) = A\,u.$$

Then the solution of the Riemann problem can be written as

$$\begin{aligned} u^*(\bar{u}_j, \bar{u}_{j+1}) &= \bar{u}_j + \sum_{\lambda_k<0} \alpha_k^{(j+1/2)} r_k \\ &= \bar{u}_{j+1} - \sum_{\lambda_k>0} \alpha_k^{(j+1/2)} r_k, \end{aligned}$$

since the contribution to the jump with $\lambda_k < 0$ will propagate to the left, and the contribution with $\lambda_k > 0$ will propagate to the right.

Substituting this expression into (2.8) one has

$$\begin{aligned} \bar{u}_j^{n+1} &= \bar{u}_j^n - \frac{\Delta t}{\Delta x}\left[A\left(\bar{u}_j + \sum_{\lambda_k<0} \alpha_k^{(j+1/2)} r_k\right) - A\left(\bar{u}_j + \sum_{\lambda_k>0} \alpha_k^{(j-1/2)} r_k\right)\right] \\ &= \bar{u}_j^n - \frac{\Delta t}{\Delta x}\left[\sum_{\lambda_k<0} \alpha_k^{(j+1/2)} \lambda_k r_k + \sum_{\lambda_k>0} \alpha_k^{(j-1/2)} \lambda_k r_k\right]. \end{aligned}$$

Multiplying by Q^{-1}, and considering that $Q^{-1} r_k = e_k$, one has

$$v_j^{n+1} = v_j^n - \frac{\Delta t}{\Delta x}\left[\sum_{\lambda_k<0} \lambda_k (v_{j+1}^{(k)} - v_j^{(k)}) e_k + \sum_{\lambda_k>0} \lambda_k (v_j^{(k)} - v_{j-1}^{(k)}) e_k\right],$$

where $\bar{u}_j^n = Q\,v_j^n$, v_j being characteristic variables, $\alpha_k^{(j+1/2)} = v_{j+1}^{(k)} - v_j^{(k)}$, and e_k is the k-th column of the $m_c \times m_c$ identity matrix. This relation can be written as

$$v_j^{n+1} = v_j^n - \frac{\Delta t}{\Delta x}\left(\Lambda_-(v_{j+1} - v_j) + \Lambda_+(v_j - v_{j-1})\right),$$

which is the same as Equation (2.5).

The Godunov scheme is therefore first-order accurate in space and time.

The Godunov scheme satisfies a discrete entropy inequality (assuming the original system satisfies an entropy inequality), namely

$$\eta(\bar{u}_j^{n+1}) \le \eta(\bar{u}_j^n) - \frac{k}{h}\left[\Psi(\bar{u}_j, \bar{u}_{j+1}) - \Psi(\bar{u}_{j-1}, \bar{u}_j)\right], \tag{2.9}$$

with $\Psi(\bar{u}_j, \bar{u}_{j+1}) = \psi(u^*(\bar{u}_j, \bar{u}_{j+1}))$.

In order to prove this, let us start from the entropy inequality of the original system:

$$\frac{\partial \eta}{\partial t} + \frac{\partial \psi}{\partial x} \le 0.$$

Let us integrate this over cell $I_j \times [t_n, t_{n+1}]$:

$$\begin{aligned}
\langle \eta((u)) \rangle_j^{n+1} &\le \langle \eta((u)) \rangle_j^n - \frac{1}{h}\left[\int_{t_n}^{t_{n+1}} \psi(u(x_{j+1/2}, t))dt - \int_{t_n}^{t_{n+1}} \psi(u(x_{j-1/2}, t))dt\right] \\
&= \eta(\bar{u}_j^n) - \frac{1}{h}\int_{t_n}^{t_{n+1}} \left[\psi(u^*(\bar{u}_j, \bar{u}_{j+1})) - \psi(u^*(\bar{u}_{j-1}, \bar{u}_j))\right] dt \\
&= \eta(\bar{u}_j^n) - \frac{k}{h}\left[\psi(u^*(\bar{u}_j, \bar{u}_{j+1})) - \psi(u^*(\bar{u}_{j-1}, \bar{u}_j))\right].
\end{aligned}$$

where

$$\langle \eta(u) \rangle_j^k \equiv \frac{1}{h}\int_{I_j} \eta(u(x, t_k))\, dx.$$

Here we made use of the fact that $u(x, t_n)$ is constant in all I_j, therefore $\langle \eta(u) \rangle_j^n = \eta(\bar{u}_j^n)$. Now, because the function η is convex (by definition of entropy), one has that $\eta(\bar{u}_j) \le \langle \eta(u) \rangle_j$ (Jensen's inequality) and therefore

$$\eta(\bar{u}_j^{n+1}) = \eta\left(\frac{1}{h}\int_{I_j} u(x, t^{n+1})dx\right) \le \frac{1}{h}\int_{I_j} \eta(u(x, t^{n+1}))dx.$$

From these two inequalities the discrete entropy inequality (2.9) follows. In addition to the entropy condition, the Godunov method, when applied to a scalar equation, has several nice properties. In particular, it is a monotone scheme (see discussion later).

The Godunov method is based on the solution of the Riemann problem, which makes it expensive to use in many circumstances. Several approximate Riemann solvers have been developed, that make Godunov methods more efficient. The most popular is the one derived by Phil Roe [53]. Another approximate Riemann solver has been proposed by Harten, Lax and Van Leer [26]. We shall not describe exact or approximate Riemann solvers. See, for example, the books by LeVeque [38, 39].

Chapter 3

The Numerical Flux Function

3.1 Higher-Order Extensions of the Godunov Method

They can be obtained by several techniques. One is to use a more accurate reconstruction of the function from cell averages, such as, for example, a piecewise linear function, and then solve the generalized Riemann problem (see Figure 3.1)

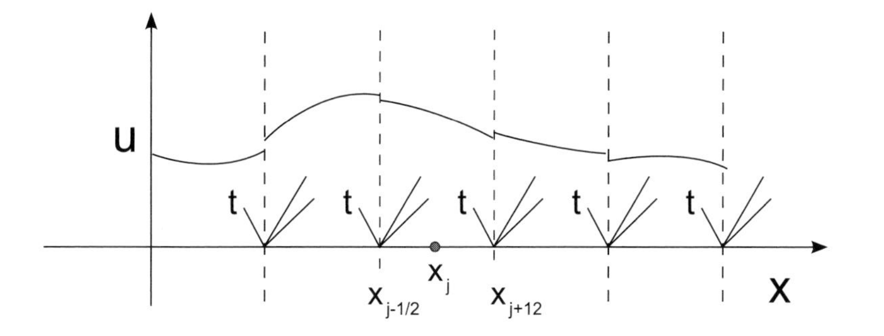

Figure 3.1: Piecewise smooth reconstruction, generalized Riemann problem and high-order extension of the Godunov method.

This approach has been used, for example, by Van Leer.

A second alternative is to use a semidiscrete scheme of the form

$$\frac{d\bar{u}_j}{dt} = -\frac{F(u^-_{j+1/2}, u^+_{j+1/2}) - F(u^-_{j-1/2}, u^+_{j-1/2})}{\Delta x}, \tag{3.1}$$

where $F(u^-, u^+)$ can be, for example, the flux function defining a Godunov scheme $F(u^-, u^+) = f(u^*(u^-, u^+))$, or some other numerical flux function, and the values $u^+_{j+1/2}$, $u^-_{j+1/2}$ are obtained by a suitable reconstruction from cell averages. Because of the relevance of this aspect, a section will be devoted to the reconstruction later. Now we shall concentrate on the properties of the numerical flux function.

3.2 The Scalar Equation and Monotone Fluxes

When dealing with nonlinear problems, linear stability is usually not enough to ensure that a numerical solution converges to a function (and therefore to a weak solution of the conservation law). The upwind scheme

$$u_j^{n+1} = u_j^n - \lambda(u_j^n - u_{j-1}^n)$$

is stable in the L^1-norm, as it is easy to check for the scalar equation, provided $0 \leq \lambda \leq 1$. The same can be said about the Lax-Friedrichs scheme, while the Lax-Wendroff scheme, on the contrary, is not stable in the L^1-norm, since the coefficients that appear in the three-point formula are not all positive (the L^1-stability is strictly related to the positivity of the coefficients, as we shall see).

To prove that, under the assumption that CFL condition is satisfied, i.e., that $\left|f'(u)\frac{\Delta t}{\Delta x}\right| < 1$, the Lax-Friedrichs scheme is L^1-stable for the equation

$$u_t + f(u)_x = 0,$$

one can act as follows: let us assume periodic boundary conditions, and let u_j^n, v_j^n be two numerical solutions.

$$\begin{aligned} u_j^{n+1} &= \frac{1}{2}(u_{j+1}^n + u_{j-1}^n) - \frac{\Delta t}{2\Delta x}(f(u_{j+1}^n) - f(u_{j-1}^n)), \\ v_j^{n+1} &= \frac{1}{2}(v_{j+1}^n + v_{j-1}^n) - \frac{\Delta t}{2\Delta x}(f(v_{j+1}^n) - f(v_{j-1}^n)). \end{aligned}$$

Let us take the difference:

$$\begin{aligned} u_j^{n+1} - v_j^{n+1} &= \frac{1}{2}(u_{j+1}^n - v_{j+1}^n) - \frac{\Delta t}{2\Delta x}(f(u_{j+1}^n) - f(v_{j+1}^n)) \\ &\quad + \frac{1}{2}(u_{j-1}^n - v_{j-1}^n) + \frac{\Delta t}{2\Delta x}(f(u_{j-1}^n) - f(v_{j-1}^n)). \end{aligned}$$

By Lagrange's theorem of the mean, one has

$$f(u_j^n) - f(v_j^n) = f'(\xi_j)(u_j^n - v_j^n),$$

where ξ_j is between u_j^n and v_j^n. Using this relation, one has

$$u_j^{n+1} - v_j^{n+1} = \frac{1}{2}(1 - \lambda_{j+1})(u_{j+1}^n - v_{j+1}^n) + \frac{1}{2}(1 + \lambda_{j-1})(u_{j-1}^n - v_{j-1}^n),$$

where

$$\lambda_j \equiv \frac{\Delta t}{\Delta x} f'(\xi_j).$$

Summing the absolute values of both sides, one has:

$$
\begin{aligned}
\|u^{n+1} - v^{n+1}\|_{L^1} &= \sum_{j=1}^{J} |u_j^{n+1} - v_j^{n+1}| \\
&= \frac{1}{2}\sum_{j=1}^{J} |(1-\lambda_{j+1})(u_{j+1}^n - v_{j+1}^n) + (1+\lambda_{j-1})(u_{j-1}^n - v_{j-1}^n)| \\
&\le \frac{1}{2}\sum_{j=1}^{J} \big(|1-\lambda_{j+1}|\,|u_{j+1}^n - v_{j+1}^n| + |1+\lambda_{j-1}|\,|u_{j-1}^n - v_{j-1}^n|\big) \\
&= \frac{1}{2}\sum_{j=1}^{J} (1-\lambda_{j+1})|u_{j+1}^n - v_{j+1}^n| + (1+\lambda_{j-1})|u_{j-1}^n - v_{j-1}^n| \\
&\text{(since } |\lambda_j| < 1) \\
&= \frac{1}{2}\sum_{j=1}^{J} (1-\lambda_j)|u_j^n - v_j^n| + (1+\lambda_j)|u_j^n - v_j^n)| \\
&\text{(because of periodicity)} \\
&= \frac{1}{2}\sum_{j=1}^{J} |u_j^n - v_j^n| = \|u_j^n - v_j^n\|_{L^1}\,.
\end{aligned}
$$

A similar proof can be given of the upwind scheme, under the hypothesis

$$0 < \frac{\Delta t}{\Delta x} f'(u) < 1\,.$$

What is the meaning of this L^1-contraction property?

When studying the stability properties of some numerical method, one usually checks whether, for the numerical solution, the same stability properties of the analytical solution hold. Now, for an entropic solution of a scalar conservation law, of the form

$$\frac{\partial u}{\partial t} + \frac{\partial f}{\partial x} = 0\,,$$

satisfying

$$\frac{\partial \eta}{\partial t} + \frac{\partial \psi}{\partial x} \le 0\,,$$

for each entropy-entropy flux (η, ψ), with convex $\eta(u)$, the following properties hold:

1. *Monotonicity preservation.* Consider the equation $u_t + f(u)_x = 0$ on the whole real line, suppose that the initial profile $u(x, 0) = u_0(x)$ is monotone, and assume that

$$\lim_{x\to\pm\infty} u_0(x) = u_{\pm\infty}$$

is bounded. Then $TV(u_0) = |u_{+\infty} - u_{-\infty}|$.

Any non-monotone profile with the same asymptotic values would have a larger total variation.

As a corollary of the TVD property of weak solutions of the scalar equation, one has that the solution $u(x,t)$ preserves monotonicity, i.e., if the initial profile is monotone, then $u(\cdot,t)$ will be monotone.

2. *TVD (Total Variation Diminishing) property.*

$$TV(u(\cdot,t_2)) \leq TV(u(\cdot,t_1)) \quad \forall t_2 \geq t_1,$$

where for all $v \in L^1(a,b)$, $TV(v)$ denotes the total variation.

We recall that the total variation of a function of a real variable is defined as

$$TV(u) \equiv \sup_{\xi} \sum_{j=1}^{N} |v(\xi_{j-1}) - v(\xi_j)|$$

where the sup is taken over all possible subdivisions $\{\xi_0, \xi_1, \ldots, \xi_N\}$ of $[a,b]$.

The total variation of a function is a measure of its oscillatory behavior. TVD means that the total amount of oscillations decreases in time. The interval $[a,b]$ may be periodic or the whole real line.

3. *L^1-contraction.* Any weak solution of a scalar conservation satisfies

$$\|u(\cdot,t_2)\|_1 \leq \|u(\cdot,t_1)\|, \quad \forall t_2 \geq t_1$$

and, more generally, given two solutions u, v, with initial condition u_0 and v_0, such that $u_0 - v_0$ has compact support, one has that $\|u(\cdot,t) - v(\cdot,t)\|_1$ is a non-increasing function of time.

4. Monotonicity.
Any pair of weak solutions $u(x,t)$, $v(x,t)$ with

$$v_0(x) \geq u_0(x) \quad \forall x,$$

satisfies

$$v(x,t) \geq u(x,t) \quad \forall x,t.$$

In constructing numerical schemes for conservation laws one tries to preserve some of such properties. In particular, a numerical scheme for scalar conservation law is said

1. *Monotonicity preserving* if

$$u_j^0 \geq u_{j+1}^0 \quad \forall j \quad \Rightarrow \quad u_j^n \geq u_{j+1}^n \quad \forall n,j,$$

$$u_j^0 \leq u_{j+1}^0 \quad \forall j \quad \Rightarrow \quad u_j^n \leq u_{j+1}^n \quad \forall n,j.$$

Just as in the continuous case, TVD implies monotonicity preservation, therefore, if a scheme is TVD, it is also monotonicity preserving.

2. *TVD* if $TV(u^{n+1}) \leq TV(u^n)$.

   In the discrete case, one has $TV(u^n) = \sum_j |u^n_{j+1} - u^n_j|$. (We assume periodic boundary conditions. Minor modifications are necessary with other boundary conditions).

3. L^1-*contractive* if
$$\|u^{n+1} - v^{n+1}\|_1 \leq \|u^n - v^n\|_1 .$$

   First-order upwind and Lax-Friedrichs schemes have this property, as we have proved.

   L^1-contraction implies TVD. In fact, given $u^n = \{u^n_j\}$, define $v^n = \{v^n_j\}$ as $v^n_j = u^n_{j-1}$. Then, if a scheme is L^1-contracting,
$$\begin{aligned} TV(u^{n+1}) &= \sum_j |u^{n+1}_j - u^{n+1}_{j-1}| = \sum_j |u^{n+1}_j - v^{n+1}_j| \\ &\leq \sum_j |u^n_j - v^n_j| = \sum_j |u^n_j - u^n_{j-1}| = TV(u^n). \end{aligned}$$

4. *Monotone* if the discrete analogue of the monotone property holds, i.e., if
$$v^n_j \geq u^n_j \quad \Rightarrow \quad v^{n+1}_j \geq u^{n+1}_j .$$

   A very simple sufficient condition to prove monotonicity is the following. Let a method be defined by an iteration function H:
$$u^{n+1}_j = H_j(u^n).$$

   Then, if for all j, $H_j(u)$ is a non-decreasing function of all arguments, then it is clear that
$$v^n_j \geq u^n_j \quad \Rightarrow \quad H_j(v^n) \geq H_j(u^n),$$
   and therefore the scheme is monotone.

   It can be proved that a monotone method is also L^1-contracting (and therefore TVD).

   It is easy to prove that the Lax-Friedrichs method is monotone.

   Although condition $\partial H_i / \partial u_j \geq 0$ for all i, j is easy to verify, monotone schemes suffer a serious restriction: it can be proved that a monotone method is at most first-order accurate. For this reason one usually looks for TVD scheme which are not monotone, and that can be higher-order accurate.

Nevertheless, the notion of *monotone schemes*, and in particular of *monotone numerical flux*, is a very important concept in the derivation of high-order schemes. A monotone flux is a flux associated to a monotone method.

Let us consider a numerical flux of the form

$$F_{j+1/2} = F(u_j, u_{j+1}).$$

We show that if a suitable CFL condition is satisfied and if F is non-decreasing in the first argument and non-increasing in the second argument (symbolically $F(\uparrow, \downarrow)$), then the corresponding scheme is monotone.

It is in fact

$$H_j(u) = u_j - \frac{k}{h}\left(F(u_j, u_{j+1}) - F(u_{j-1}, u_j)\right),$$

$$\begin{aligned}
\frac{\partial H_j}{\partial u_j} &= 1 - \frac{k}{h}\left(\frac{\partial F}{\partial u^{(1)}}(u_j, u_{j+1}) - \frac{\partial F}{\partial u^{(2)}}(u_{j-1}, u_j)\right), \qquad (3.2)\\
\frac{\partial H_j}{\partial u_{j+1}} &= -\frac{k}{h}\frac{\partial F}{\partial u^{(2)}}(u_j, u_{j+1}) \geq 0,\\
\frac{\partial H_j}{\partial u_{j-1}} &= \frac{k}{h}\frac{\partial F}{\partial u^{(1)}}(u_j, u_{j+1}) \geq 0,\\
\frac{\partial H_j}{\partial u_l} &= 0, \quad l \neq j-1, j, j+1.
\end{aligned}$$

Positivity for (3.2) has to be checked separately.

For example, both upwind and Lax-Friedrichs flux satisfy condition $F(\uparrow, \downarrow)$. For the first scheme, condition $\partial H_j/\partial u_j \geq 0$ coincides with the CFL condition, while for the Lax-Friedrichs scheme $\partial H_j/\partial u_j = 0$, and the other two conditions correspond to the CFL condition. Other popular numerical flux functions that satisfy the $F(\uparrow, \downarrow)$ condition are:

i) Godunov flux:

$$F(u,v) = \begin{cases} \min_{u \leq \xi \leq v} f(\xi) & \text{if } u \leq v, \\ \max_{u \leq \xi \leq v} f(\xi) & \text{if } u > v. \end{cases} \qquad (3.3)$$

ii) Engquist-Osher flux [18]:

$$F(u,v) = \int_0^u \max(f'(\xi), 0)\, d\xi + \int_0^u \min(f'(\xi), 0)\, d\xi + f(0).$$

iii) Local Lax-Friedrichs flux (also called Rusanov flux):

$$F(u,v) = \frac{1}{2}(f(u) + f(v) + \alpha(u - v)),$$

where $\alpha = \max_w |f'(w)|$, the maximum being taken over the relevant range of u.

Several other numerical flux functions can be obtained by approximate Riemann solvers, such as the Roe solver [53] of the HLL solver [26]. For an account of several approximate solvers and numerical flux functions see, for example, the book by LeVeque [39].

Remark 3.1. The flux functions are listed in increasing order of dissipativity. Although all fluxes can be used to construct a first-order monotone scheme, not all fluxes give the some numerical results. Godunov flux will provide shaper discontinuities than Local Lax-Friedrichs.

Remark 3.2. Local Lax-Friedrichs differ froms Lax-Friedrichs flux by the fact that the amount of dissipation is different for each cell interface, and is taken to be the minimum possible compatible with stability. LLxF is much less dissipative than standard LxF.

Remark 3.3. Monotone fluxes are very important as essential building blocks in the construction of high-order finite-volume methods. As we shall see, the non-oscillatory properties of a scheme depend on the numerical flux and on the non-oscillatory reconstruction.

Remark 3.4. All these concepts have been for the scalar equation. They will be used as guidelines in the development of numerical schemes that will be used for systems as well.

Chapter 4

Nonlinear Reconstruction and High-Order Schemes

4.1 High-Order Finite-Volume Schemes

General structure of high-order finite-volume schemes. In a finite-volume scheme, the basic unknown is the cell average $\bar{u}_j$.

We have seen that the solution $u(x,t)$ satisfies the equation

$$\frac{d < u >_j}{dt} + \frac{f(u(x_{j+1/2},t)) - f(u(x_{j-1/2},t))}{h} = 0,$$

where $< u >_j \equiv \frac{1}{h}\int_{I_j} u(x,t)\, dt$.

First-order (in space) semidiscrete schemes can be obtained using $F_{j+1/2} = F(\bar{u}_j, \bar{u}_{j+1})$ in place of $f(u(x_{j+1/2},t))$:

$$\frac{d\bar{u}_j}{dt} = -\frac{F(\bar{u}_j, \bar{u}_{j+1}) - F(\bar{u}_{j-1}, \bar{u}_j)}{h}.$$

A scheme based on this formula, however, is restricted to first-order accuracy.

Higher-order schemes are obtained by using a piecewise polynomial reconstruction in each cell, and evaluating the numerical flux on the two sides of the interface (see Figure 4.1):

$$\frac{d\bar{u}_j}{dt} = -\frac{F(u^-_{j+1/2}, u^+_{j+1/2}) - F(u^-_{j-1/2}, u^+_{j-1/2})}{h}.$$

A second-order scheme is obtained by a piecewise linear reconstruction. This is obtained as follows.

Given $\{u^n_j\}$, compute a piecewise linear reconstruction

$$L(x) = \sum_j L_j(x)\chi_j(x)$$

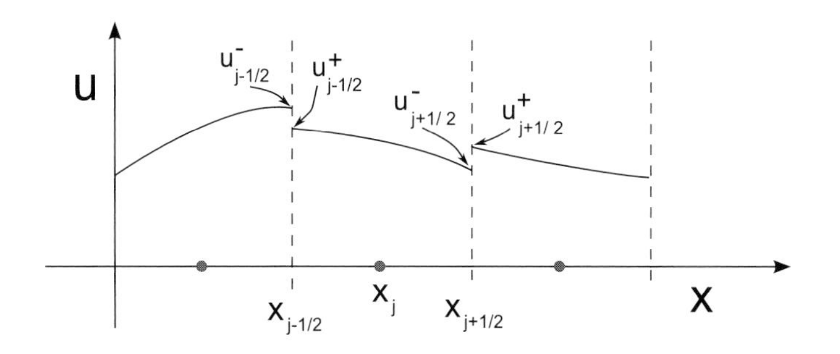

Figure 4.1: Function reconstruction at cell edges.

with $L_j = \bar{u}_j + u'_j(x - x_j)$. The quantity u'_j is a suitable (first-order) approximation of the space derivative of the profile $u(x)$ at x_j.

The numerical approximation of the first derivative is a very important point, since the accuracy and TVD properties of the scheme depend on it. If, for example, one uses standard central difference, then the reconstructed piecewise linear function will have spurious extrema, and its total variation will be larger that the total variation of the discrete data.

In order to prevent the formation of spurious extrema, the derivative has to be reconstructed by a suitable *limiter*. The simplest one is the so called *minmod* limiter, defined as

$$\mathrm{MinMod}(a,b) = \begin{cases} a & \text{if} \quad |a| \le |b| \quad \text{and} \quad ab > 0, \\ b & \text{if} \quad |a| > |b| \quad \text{and} \quad ab > 0, \\ 0 & \text{if} \quad ab \le 0. \end{cases} \tag{4.1}$$

The effect of the MinMod limiter is illustrated in Figure 4.2. The minmod limiter

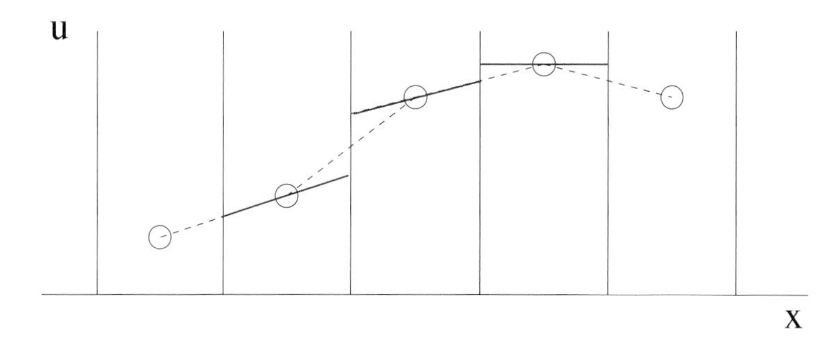

Figure 4.2: Slope reconstruction using the MinMod limiter.

is very robust, but has the drawback of degrading the accuracy of the scheme to first-order near local extrema.

More accurate limiter strategies are described, for example, in [38].

Once the profile is reconstructed, then the function at the edge of the cell is given by

$$u_{j+1/2}^- = L_j(x_{j+1/2}),$$
$$u_{j+1/2}^+ = L_{j+1}(x_{j+1/2}).$$

4.2 Essentially Non-Oscillatory Reconstruction (ENO)

High-order reconstruction has the purpose of providing higher accuracy in space. This step is a crucial one in the development of shock-capturing schemes, since naive reconstructions may introduce spurious oscillations in the profile. In this and in the next section we shall describe with some detail two of the major techniques commonly used to prevent formation of (large) spurious oscillations, still guaranteeing high-order accuracy in smooth regions. First we start with the ENO reconstruction.

The goal is the following. Let us assume there is a smooth function $u(x)$, and we know only its cell averages $\{\bar{u}_j\}$. Then we want to construct in each cell j a polynomial p_j of a given degree $m-1$ (i.e., $p_j \in \prod_{m-1}$)

$$p_j(x) = u(x) + O(\Delta x^m). \tag{4.2}$$

In particular, we shall be interested in evaluating this polynomial at cell boundaries:

$$u_{j+1/2}^- = p_j(x_{j+1/2}),$$
$$u_{j-1/2}^+ = p_j(x_{j-1/2}).$$

Such a polynomial is constructed as follows. Take m adjacent cells, that include cell j. Let these cells be denoted by $j-r, j-r+1, \ldots, j+s$ with $r+s+1=m$, $r,s \geq 0$. Then impose that

$$< p_j >_l = < u >_l = \frac{1}{h}\int_{x_{l-1/2}}^{x_{l+1/2}} u(x)\,dx,\ l = j-r,\ldots,j+s. \tag{4.3}$$

These m independent conditions uniquely determine a polynomial of degree $m-1$. Let us show that indeed this polynomial satisfies condition (4.2). This is easily shown by defining

$$U(x) \equiv \int_a^x u(\tilde{x})\,d\tilde{x},$$

a primitive of $u(x)$. The left bound a is not relevant. We choose it so that $a = x_{j_a-1/2}$. At the right edge of cell i one has:

$$U(x_{i+1/2}) = \Delta x \sum_{j=j_a}^{i} \bar{u}_j.$$

Let $P_j(x) \in \prod_m$, and let $P_j(x_{i+1/2}) = U(x_{i+1/2})$, $i = j-r-1,\dots,j+s$. These $m+1$ conditions uniquely determine $P_j \in \prod_m$. Furthermore, from interpolation theory, one has

$$P(x) = U(x) + O(\Delta x^{m+1}),$$

and therefore

$$p(x) = P'(x) = U'(x) + O(\Delta x^m) = u(x) + (\Delta x^m).$$

The polynomial $p(x)$ therefore satisfies (4.2) and (4.3).

There are m such polynomials. For example, for a polynomial of degree 2 one can choose cells $j-2, j-1, j$ or $j-1, j, j+1$, or $j, j+1, j+2$. Which one should one choose for the reconstruction? This is exactly where ENO comes into play. First, observe that for a given stencil, the polynomial $P(x)$ can be computed using the divided difference of the function $U(x)$:

$$U[x_{i-1/2}, x_{i+1/2}] = \frac{U(x_{i+1/2}) - U(x_{i-1/2})}{x_{i+1/2} - x_{i-1/2}} = \bar{u}_i,$$

therefore first- and higher-order divided differences of U can be computed by $\{\bar{u}_j\}$, without using function U explicitly. Likewise, computation of $p(x)$ can be performed using divided differences that make use only of $\{\bar{u}_j\}$. The main purpose of the primitive function $U(x)$ is to find the proper stencil.

The idea of ENO construction is the following. Take cell j, and construct a linear function between point $(x_{j-1/2}, U(x_{j-1/2}))$ and $(x_{j+1/2}, U(x_{j+1/2}))$. Let us call it $P_1(x)$. Then add one point, either to the left obtaining

$$R(x) = P_1(x) + U[x_{j-3/2}, x_{j-1/2}, x_{j+1/2}](x - x_{j-1/2})(x - x_{j+1/2}),$$

or on the right, obtaining

$$R(x) = P_1(x) + U[x_{j-1/2}, x_{j+1/2}, x_{j+3/2}](x - x_{j-1/2})(x - x_{j+1/2}),$$

Then chose the one which is less oscillatory, i.e., the one with the smallest second derivative. Therefore, extend your stencil by:

$$r \to r+1 \ \text{ if } \ |U[x_{j-3/2}, x_{j-1/2}, x_{j+1/2}]| < |U[x_{j-1/2}, x_{j+1/2}, x_{j+3/2}]|,$$

or

$$s \to s+1 \ \text{ if } \ |U[x_{j-3/2}, x_{j-1/2}, x_{j+1/2}]| > |U[x_{j-1/2}, x_{j+1/2}, x_{j+3/2}]|.$$

Then one can repeat the procedure by adding one more point to the stencil, either to the left or to the right, comparing the size of the next divided difference. Suppose for example that the function has a discontinuity across a cell boundary (see Figure 4.3). The reconstruction in the upper part of the figure is obtained by using, in each cell i, the parabola obtained by matching the cell averages in

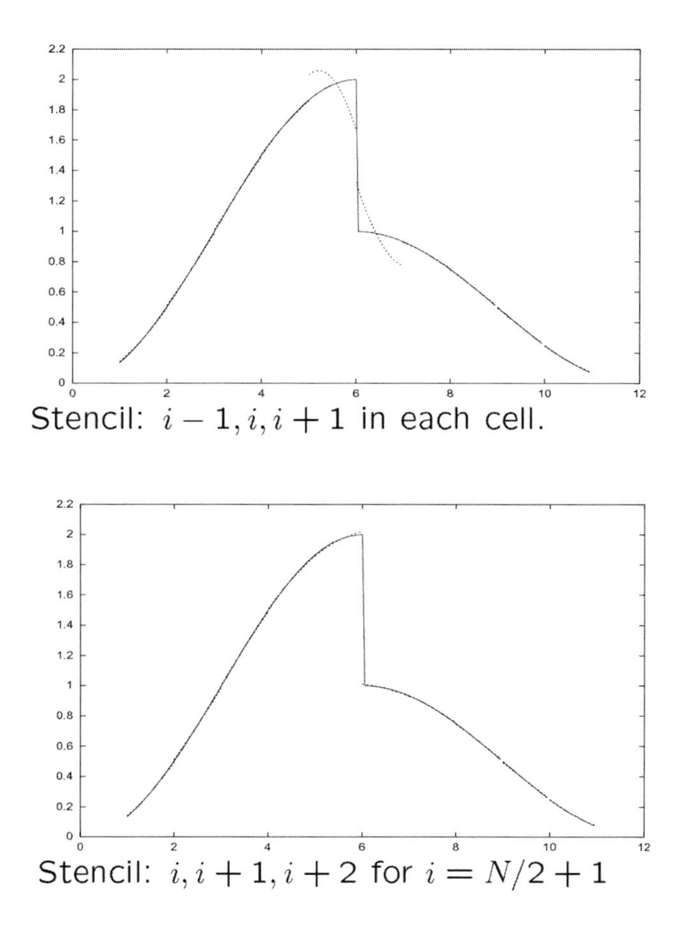

Figure 4.3: Piecewise parabolic reconstruction of a piecewise smooth profile. Upper: central reconstruction in each cell; lower: ENO reconstruction.

cells $i-1$, i, and $i+1$, while the one in the lower part is obtained by ENO. The procedure chooses, for example, stencil based on cells i, $i+1$ and $i+2$ for the cell just on the right of the discontinuity (see Figure 4.3)

The net effect of this procedure will be to choose a stencil that uses the smooth part of the function in the reconstruction.

Remark 4.1. For a given degree $m-1$, there are m possible stencils. For each of them there are two sets of coefficients, $\{c_{ri}\}$, $\{\tilde{c}_{ri}\}$ that compute $u^-_{j+1/2}$ and $u^+_{j-1/2}$ as a linear combination of cell averages on the stencil. These coefficients can be computed once and used later. Once the stencil is chosen (i.e., r is defined) by the ENO procedure, then one knows which set of coefficients c_{ri} one has to use. The expressions for $u^-_{j+1/2}$ and $u^+_{j-1/2}$ for each choice of the stencil are of the form

$$u^-_{j+1/2} = \sum_{i=0}^{m-1} c_{r\,i}\bar{u}_{j-r+i}\,,$$

$$u^+_{j-1/2} = \sum_{i=0}^{m-1} \tilde{c}_{r\,i}\bar{u}_{j-r+i}\,.$$

Remark 4.2. In the choice of the stencil it is better to prefer more centered stencil ($r \approx s$) for comparable values of the divided difference, since the interpolation error is lower. This effect can be taken into account by a bias in the choice of the stencil toward centered stencils.

Remark 4.3. For each fixed stencil, one obtains a different scheme. If one performs a linear stability analysis of such schemes, one will see that some of them are unstable. However, this is not in general a problem. Even if the ENO procedure occasionally chooses a linearly unstable scheme, the choice is usually temporary, and lasts for one or few time steps, preventing the linearly unstable mode to amplify substantially.

Remark 4.4. Piecewise polynomial reconstruction is a convenient choice because the polynomials can be easily reconstructed by the Newton procedure. However, other functions could be used as a basis for the piecewise (essentially) non-oscillatory reconstruction. Marquina showed that the use of hyperbolas in the reconstruction produces a less oscillatory profile than the use of parabolas [43]. Furthermore, hyperbolic reconstructions have also the advantage of reducing dissipation when local Lax-Friedrichs flux is used.

4.3 Weighted ENO Reconstruction (WENO)

In the ENO reconstruction one chooses a stencil with m nodes to construct a polynomial of degree $m-1$, in order to reach an accuracy of $O(h^m)$ in the cell. However, the total number of points involved is $2m-1$. With all these points, a much more accurate reconstruction is possible. Suppose, for the sake of argument, that $m=3$, and that we use a parabola to reconstruct the function $u(x)$ in cell j. Let q_k denote the parabola obtained by matching the cell average in cells $k-1$, k, $k+1$, i.e., $q_k(x)$ is obtained by imposing

$$< q_k >_l = \bar{u}_l\,, \quad l = k-1, k, k+1\,.$$

Then for our polynomial $p_j \in \prod_2$ we can use either q_{j-1}, q_j, or q_{j+1}. Each choice would give us third-order accuracy. We could also choose a convex combination of q_k,

$$p_j = w^j_{-1}q_{j-1} + w^j_0 q_j + w^j_1 q_{j+1}\,,$$

with $w_{-1}^j + w_0^j + w_1^j = 1$, $w_l^j \geq 0$, $l = -1, 0, 1$. Every such convex combination would provide at least third-order accuracy.

We can choose the weights according to the following requirements:

i) in the region of regularity of $u(x)$ the values of the weights are chosen in such a way to have a reconstruction of the function at some particular point with higher order of accuracy. Typically we need high-order accuracy at points $x_j + \frac{h}{2}$ and $x_j - \frac{h}{2}$. With two more degrees of freedom it is possible to obtain fifth-order accuracy at point $x_{j+1/2}$ (instead of third-order).

   We shall denote by C_{-1}^+, C_0^+, C_1^+ the constants that provide high-order accuracy at point $x_{j+1/2}$:

$$p_j(x_{j+1/2}) = \sum_{k=-1}^{1} C_k^+ q_{j+k}(x_{j+1/2}) = u(x_{j+1/2}) + O(h^5),$$

   and C_k^-, $k = -1, 0, 1$ the corresponding constants for high-order reconstruction at point $x_{j-1/2}$.

$$p_j(x_{j-1/2}) = \sum_{k=-1}^{1} C_k^- q_{j+k}(x_{j-1/2}) = u(x_{j-1/2}) + O(h^5).$$

   The values of these constants can be computed, and are given by

$$C_1^+ = C_{-1}^- = \frac{3}{10}, \quad C_0^+ = C_0^- = \frac{3}{5}, \quad C_{-1}^+ = C_1^- = \frac{1}{10}.$$

ii) In the region near a discontinuity, one should make use only of the values of the cell averages that belong to the regular part of the profile.
   Let us consider the example in Figure 4.4.

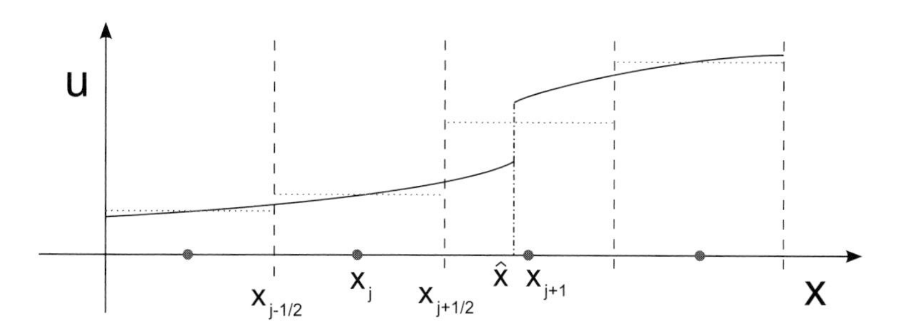

Figure 4.4: Discontinuous profile and weight selection in WENO scheme.

Suppose that the function $u(x)$ has a discontinuity in $\hat{x} \in I_{j+1}$. Then in order to reconstruct the function in cell j one would like to make use only of q_{j-1}, i.e., the weights should be

$$w_{-1}^j \sim 1, \quad w_0^j \sim 0, \quad w_1^j \sim 0.$$

This is obtained by making the weights depend on the regularity of the function in the corresponding cell. In usual WENO scheme this is obtained by setting

$$\alpha_k^j = \frac{C_k}{(\beta_k^j + \epsilon)^2}, \qquad k = -1, 0, 1,$$

and

$$w_k^j = \frac{\alpha_k^j}{\sum_\ell \alpha_\ell^j}.$$

Here β_k are the so-called *smoothness indicators*, and are used to measure the smoothness or, more precisely, the roughness of the function, by measuring some weighted norm of the function and its derivatives. Typically,

$$\beta_k^j = \sum_{l=1}^{2} \int_{x_{j-1/2}}^{x_{j+1/2}} h^{2l-1} \left(\frac{d^l q_{j+k}(x)}{dx^l} \right)^2 dx, k = -1, 0, 1.$$

The integration can be carried out explicitly, obtaining

$$\begin{aligned} \beta_{-1} &= \frac{13}{12}(\bar{u}_{j-2} - 2\bar{u}_{j-1} + \bar{u}_j)^2 + \frac{1}{4}(\bar{u}_{j-2} - 4\bar{u}_{j-1} + 3\bar{u}_j)^2, \\ \beta_0 &= \frac{13}{12}(\bar{u}_{j-1} - 2\bar{u}_j + \bar{u}_{j+1})^2 + \frac{1}{4}(\bar{u}_{j-1} - \bar{u}_{j+1})^2, \\ \beta_1 &= \frac{13}{12}(\bar{u}_j - 2\bar{u}_{j+1} + \bar{u}_{j+2})^2 + \frac{1}{4}(3\bar{u}_j - 4\bar{u}_{j+1} + \bar{u}_{j+2})^2. \end{aligned}$$

With three parabolas one obtains a reconstruction that gives up to fifth-order accuracy in a smooth region, and that degrades to third-order near discontinuities.

A detailed account of ENO and WENO reconstruction as well as of finite-volume and finite-difference schemes can be found in Chapter 4 of [56], written by C.W. Shu for the CIME course "Advanced Numerical Approximation of Nonlinear Hyperbolic Equations", held in Cetraro in 1997. These Lecture Notes contain also a nice overview of different approaches to the numerical solution of hyperbolic systems written by E. Tadmor. Most of the material written by Shu can be downloaded from the web as ICASE Report No.97-65.

4.4 Conservative Finite-Difference Schemes

In a finite-difference scheme the basic unknown is the pointwise value of the function, rather than its cell average. Osher and Shu [58] observed that it is possible to write a finite-difference scheme in conservative form as follows. Let us consider a system

$$\frac{\partial u}{\partial t} + \frac{\partial f}{\partial x} = 0.$$

Let us write

$$\frac{\partial f}{\partial x}(u(x)) = \frac{\hat{f}(u(x+\frac{h}{2})) - \hat{f}(u(x-\frac{h}{2}))}{h}.$$

The relation between f and $\hat{f}$ is the following. Let us consider the sliding cell average operator:

$$\bar{u}(x) = \frac{1}{h}\int_{x-\frac{h}{2}}^{x+\frac{h}{2}} u(\xi)\, d\xi.$$

Differentiating with respect to x one has:

$$\frac{\partial \bar{u}}{\partial x} = \frac{1}{h}(u(x+\frac{h}{2}) - u(x-\frac{h}{2})).$$

Therefore the relation between f and $\hat{f}$ is the same that exists between $\bar{u}(x)$ and $u(x)$, namely, function f is the cell average of the function $\hat{f}$. This also suggests a way to compute the flux function. The technique that is used to compute pointwise values of $u(x)$ at the edge of the cell from cell averages of u can be used to compute $\hat{f}(u(x_{j+1/2}))$ from $f(u(x_j))$. This means that in the finite-difference method it is the flux function which is computed at x_j and then reconstructed at $x_{j+1/2}$. But the reconstruction at $x_{j+1/2}$ may be discontinuous. Which value should one use? A general answer to this question can be given if one considers flux functions that can be splitted as

$$f(u) = f^+(u) + f^-(u), \tag{4.4}$$

with the condition (for scalar fluxes) that

$$\frac{df^+(u)}{du} \geq 0, \quad \frac{df^-(u)}{du} \leq 0. \tag{4.5}$$

There is a close analogy between flux splitting and numerical flux functions. In fact, if a flux can be splitted as Equation (4.4), with property (4.5), then

$$F(a,b) = f^+(a) + f^-(b)$$

will define a monotone consistent flux.

This is the case, for example, of the local Lax-Friedrichs flux.

A finite-difference scheme therefore takes the following form:

$$\frac{du_j}{dt} = -\frac{1}{h}[\hat{F}_{j+1/2} - \hat{F}_{j-1/2}],$$

$$\hat{F}_{j+1/2} = \hat{f}^+(u^-_{j+1/2}) + \hat{f}^-(u^+_{j+1/2});$$

$\hat{f}^+(u^-_{j+1/2})$ is obtained as follows:

- compute $f^+(u_l)$ and interpret it as cell average of $\hat{f}^+$,

- perform pointwise reconstruction of $\hat{f}^+$ in cell j, and evaluate it in $x_{j+1/2}$;

$\hat{f}^-(u^+_{j+1/2})$ is obtained as follows:

- compute $f^-(u_l)$, interpret as cell average of $\hat{f}^-$,
- perform pointwise reconstruction of $\hat{f}^-$ in cell $j+1$, and evaluate it in $x_{j+1/2}$.

Remark 4.5. We used uniform meshes. Finite-volume methods can be used on arbitrary non-uniform meshes. Finite-difference can be used only on uniform (or smoothly varying) mesh. This makes finite-volume more flexible in several dimensions. They can even be constructed on unstructured grids.

Remark 4.6. For finite-volume methods applied to systems, better results are usually obtained if one uses characteristic variables rather than conservative variables in the reconstruction step. See, e.g., [14], or [50] for an example in the context of central schemes.

Remark 4.7. There is some difference in the sharpness of the resolution of the numerical results, according to the numerical flux one uses. Godunov flux gives much sharper results on linear discontinuities. The difference, however, becomes less relevant with the increase of the order of accuracy.

Remark 4.8 (Boundary conditions). A general treatment of boundary conditions for hyperbolic systems is beyond the scope of the present lecture notes. A general technique to impose simple boundary conditions consists in extending the computational domain with a given number of "ghost" cells, in which data is assigned at t_n. If one solves a system of m equations, in each ghost cell one has to assign m quantities. However, all the quantities that are assigned at the ghost cells are not independent: on the two boundaries a total of m independent quantities can be assigned, while the other m have to be compatible with the evolution equation. The characteristic condition (number of independent assigned conditions must be equal to the number of characteristic entering the domain at the boundary) has to be satisfied at a discrete level. A description of boundary conditions for a general hyperbolic system can be found, for example, in the book of Godlewski and Raviart, [21].

4.5 Time Integration: Runge-Kutta Methods

Once the system of PDE's has been reduced to a system of ODE's, it may be solved numerically by some standard ODE solver, such as, for example, Runge-Kutta.

Let us consider an initial value problem for a system of ordinary differential equations.

$$\begin{cases} y' &= g(y), \\ y(t_0) &= y_0. \end{cases}$$

Apply to the initial value problem above an explicit Runge-Kutta scheme with ν stages:

$$y^{n+1} = y^n + \Delta t \sum_{i=1}^{\nu} b_i K^{(i)}.$$

The $K^{(i)}$ are called *Runge-Kutta fluxes* and are defined by

$$K^{(i)} = g(y^{(i)}) \qquad \text{with} \quad y^{(1)} = y^n, \quad i = 1, \ldots, \nu$$

where the $y^{(i)}$ will be called *intermediate values*, and, for an explicit scheme, are given by

$$y^{(i)} = y^n + \Delta t \sum_{l=1}^{i-1} a_{i,l} K^{(l)}, \quad i = 1, \ldots, \nu - 1.$$

The matrix $A = (a_{i,l})$, and the vector b define uniquely the RK scheme. With the present notation, A is a $\nu \times \nu$ lower triangular matrix, with zero elements on the diagonal.

Implicit schemes are usually not used when source terms are not present, because hyperbolic systems are in general not stiff. The treatment of systems with stiff source will be given in Section 6.

4.6 SSP Schemes

When constructing numerical schemes for conservation laws, one has to take great care in order to avoid spurious numerical oscillations arising near discontinuities of the solution. This is avoided by a suitable choice of space discretization and time discretization.

Solution of scalar conservation equations, and equations with a dissipative source have some norm that decreases in time. It would be desirable that such property is maintained at a discrete level by the numerical method. If U^n represents a vector of solution values (for example obtained from a method of lines approach in solving (1.1) we recall the following [61]

Definition. A sequence $\{U^n\}$ is said to be *strongly stable* in a given norm $||\cdot||$ provided that $||U^{n+1}|| \leq ||U^n||$ for all $n \geq 0$.

The most commonly used norms are the TV-norm and the infinity norm. A numerical scheme that maintains strong stability at discrete level is called Strong Stability Preserving (SSP).

Here we review some basic facts about RK-SSP schemes. First, it has been shown [23] under fairly general conditions that high-order SSP schemes are necessarily explicit. Second, observe that a generic explicit RK scheme can be written as

$$
\begin{aligned}
U^{(0)} &= U^n, \\
U^{(i)} &= \sum_{k=0}^{i-1} (\alpha_{ik} U^{(k)} + \Delta t \beta_{ik} L(U^{(k)})), \quad i = 1, \dots, \nu, \\
U^{n+1} &= U^{(\nu)},
\end{aligned}
\tag{4.6}
$$

where $\alpha_{ik} \geq 0$ and $\alpha_{ik} = 0$ only if $\beta_{ik} = 0$. This representation of RK schemes (which is not unique) can be converted to a standard Butcher form in a straightforward manner. Observe that for consistency, one has $\sum_{k=0}^{i-1} \alpha_{ik} = 1$. It follows that if the scheme can be written in the form (4.6) with non-negative coefficients β_{ik}, then it is a convex combination of Forward Euler steps, with step sizes $(\beta_{ik}/\alpha_{ik})\Delta t$. A consequence of this is that if Forward Euler is SSP for $\Delta t \leq \Delta t^*$, then the RK scheme is also SSP for $\Delta t \leq c\Delta t^*$, with $c = \min_{ik}(\alpha_{ik}/\beta_{ik})$ [57, 23].

The constant c is a measure of the efficiency of the SSP-RK scheme, therefore for the applications it is important to have c as large as possible. For a detailed description of optimal SSP schemes and their properties see [61].

Some example of explicit SSP RK schemes are given in the appendix.

4.7 Extension to More Dimensions

Both finite-volume and finite-difference methods can be extended to more space dimensions, although high-order methods are considerably more expensive. As a general remark, finite-difference schemes become more efficient than finite-volume schemes in 2D, because the reconstructions can be performed direction-by-direction, while in 1D finite-volume are slightly faster, because one has to reconstruct only $u(x)$ from $\bar{u}(x)$ rather then two fluxes, $f^+(u)$ and $f^-(u)$. The treatment of systems in more dimensions goes beyond the scope of the present lecture notes. An extensive discussion of such schemes can be found, for example, in the lecture notes by Shu [56].

Chapter 5

Central Schemes

Shock-capturing finite-volume and finite-difference schemes that have been described in the previous chapters require the use of a numerical flux function, which is necessary to define the flux across the cell edges. In the Godunov representation of the solution, in fact the function $u(x, t^n)$ is piecewise smooth (in general, piecewise polynomial), with discontinuities at cell boundaries.

An alternative approach to construct high-order shock-capturing schemes is offered by the so-called *central schemes*, which are naturally (but not necessarily) constructed by making use of a staggered mesh.

As we shall see, central schemes have the attractive feature of not requiring the solution of the (exact or approximate) Riemann problem. For this reason, they are relatively simple to use on a large variety of systems. One drawback is that they are usually more dissipative than upwind-based schemes of the same order, but in general less dissipative than the corresponding non-staggered finite-volume scheme that makes use of the local Lax-Friedrichs flux.

Introducing some characteristic information in central schemes may improve their properties. More specifically, we shall see that using characteristic variables in the reconstruction will reduce spurious numerical oscillations considerably.

From the pioneering works of Nessyahu and Tadmor in [45] and Sanders and Weiser [55] an extensive literature on central schemes has developed. These central schemes are obtained integrating the conservation law in space and time on control volumes which are staggered with respect to the cells on which the cell averages are based. In this fashion, the discontinuities in the pointwise solution produced by the reconstruction algorithm are located at the center of the staggered control volumes. As a consequence, the solution is smooth at the edges of the control volumes, thus enabling a simple construction of the numerical fluxes.

High-order central schemes can be constructed by performing a reconstruction of the field from cell averages, and using this reconstruction for computing the staggered cell average, the pointwise value of the function from cell averages, and the space derivative of the flux function. High-order reconstructions can be obtained using ENO [8] or WENO [40] technique applied to central schemes.

Here we shall present a general technique to construct high-order shock-capturing central schemes on staggered grid, called Central Runge Kutta (CRK) [50]. The new approach replaces the previous high-order central schemes based on dense output for the computation of the integral of the flux along cell edges [8, 40].

Non-staggered central schemes have been developed [33], which have strong resemblance with non-staggered finite-volume schemes which make use of local Lax-Friedrichs numerical flux function.

Ideas from central and from upwind schemes can be combined, in order to obtain schemes which are less dissipative than purely central schemes, and easier to use than purely updind schems. Semidiscrete central-upwind schemes for conservation laws and for Hamilton-Jacobi equations have been derived, for example, in [34]. The approach is based on a careful estimation of the lowest and highest characteristic speeds, so only a small amount of characteristic information is used. This idea was first used by Harten, Lax and van Leer [26].

A review on central schemes can be found, for example, in [63], in which several other approaches, including spectral methods, are presented for the solution of conservation laws, and in [54] (second chapter), in which staggered central schemes with WENO reconstruction in two space dimensions are presented. The treatment of the stiff source terms in the context of central schemes is also presented. In the same book (Chapter 3) one can find description of ENO and WENO reconstruction on unstructured grids.

5.1 Nessyahu-Tadmor Second-Order Scheme

We consider the system of equations

$$u_t + f_x(u) = 0, \tag{5.1}$$

with $u \in \mathbb{R}^m$, $f : \mathbb{R}^m \to \mathbb{R}^m$ continuously differentiable. We suppose that the Jacobian of f, $A(u) = f'(u)$ has real eigenvalues and a complete set of eigenvectors.

A key point in both upwind and central schemes is the reconstruction step. From the cell averages $\{\bar{u}^n\}$, it is necessary to reconstruct the initial data $u^n(x)$. This information is needed to evaluate the right-hand side of (3.1). Typically, starting from $u^n(x)$, one computes a piecewise polynomial function of the form:

$$u^n(x) = R(x; \{\bar{u}^n\}) = \sum_j P_j^d(x)\chi_{I_j}(x), \tag{5.2}$$

where $P_j^d(x)$ is a polynomial of degree d, and χ_{I_j} is the characteristic function of interval I_j.

In central schemes based on staggered grids, the conservation law is integrated on the staggered control volume: $V_{j+1/2}^n = [x_j, x_{j+1}] \times [t^n, t^n + \Delta t]$. Integrating

(5.1) in space and time on $V^n_{j+1/2}$ and dividing by h, one finds:

$$\bar{u}^{n+1}_{j+1/2} = \bar{u}^n_{j+1/2} - \frac{1}{h}\int_0^{\Delta t} [f(u(x_{j+1}, t^n + \tau)) - f(u(x_j, t^n + \tau))\, d\tau]. \quad (5.3)$$

The first term on the right-hand side $\bar{u}^n_{j+1/2}$ is evaluated integrating exactly the reconstruction on $[x_j, x_{j+1}]$:

$$\bar{u}^n_{j+1/2} = \frac{1}{h}\int_{x_j}^{x_{j+1}} u^n(x) = \frac{1}{h}\int_{x_j}^{x_{j+1/2}} P_j(x)\, dx + \frac{1}{h}\int_{x_{j+1/2}}^{x_{j+1}} P_{j+1}(x)\, dx. \quad (5.4)$$

Since u is smooth at x_j and x_{j+1}, if the time step is small enough, the integral in time of the fluxes can be accurately evaluated through quadrature. The values of u at the nodes of the quadrature formula are predicted integrating the system of ODE's

$$\left.\frac{du}{dt}\right|_{x_j} = -\left. f_x(u)\right|_{x_j},$$

using again the smoothness of u at x_j.

An example is the second-order Nessyahu-Tadmor scheme, which is characterized by a piecewise linear reconstruction, while the fluxes are integrated with the midpoint rule, and the value of u at the midpoint is predicted with Taylor expansion. The scheme has a predictor corrector structure, and it can be described in three steps:

- **Reconstruction from cell averages**, to yield the staggered cell averages $\bar{u}^n_{j+1/2}$, and the approximate slopes u'_j

$$u^{n+1/2}_j = \frac{1}{2}(\bar{u}^n_j + \bar{u}^n_{j+1}) - \frac{1}{8}(u'_{j+1} - u'_j)h.$$

- **Predictor step.** Evaluate:

$$u^{n+1/2}_j = \bar{u}^n_j - \frac{\Delta t}{2} f'_j.$$

- **Corrector step.** Compute the new cell averages:

$$\bar{u}^{n+1}_{j+1/2} = \bar{u}^n_{j+1/2} - \lambda\left[f(u^{n+1/2}_{j+1}) - f(u^{n+1/2}_j)\right].$$

A first-order approximation of the space derivatives $f'_j \approx \partial f/\partial x(\bar{u}^n_j)$, $u'_j \approx \partial u/\partial x(x_j)$, can be computed by using a suitable slope limiter, such as MinMod or UNO (Uniformly Non Oscillatory) [25].

Higher-order schemes require the reconstruction of pointwise values of the solution from cell averages, in addition to staggered cell averages, and of space derivatives of the flux.

5.2 Description of CRK Schemes

In Central Runge Kutta schemes, the conservation law is integrated on the interval $I_{j+1/2} = [x_j, x_{j+1}]$ (see Fig.5.1). We obtain the exact equation:

$$\left.\frac{d\bar{u}}{dt}\right|_{j+1/2} = -\frac{1}{h}\left[f(u(x_{j+1},t)) - f(u(x_j,t))\right]. \tag{5.5}$$

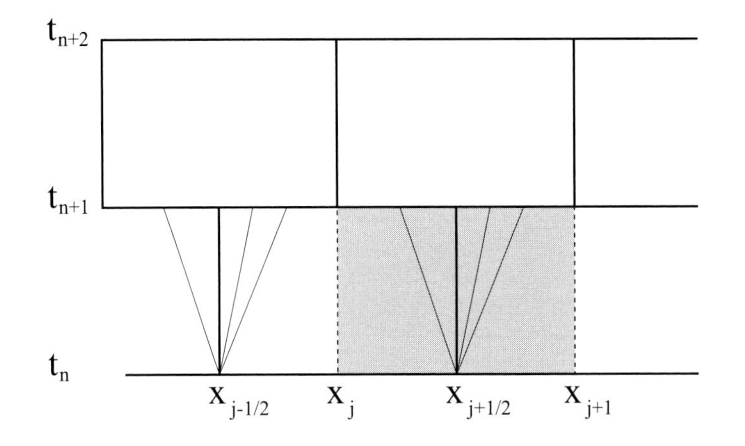

Figure 5.1: Integration over a staggered cell, and construction of central schemes.

Next, this equation is discretized in time with a Runge-Kutta scheme. Thus the updated solution will be given by:

$$\bar{u}^{n+1}_{j+1/2} = \bar{u}^n_{j+1/2} - \lambda \sum_{i=1}^{\nu} b_i K^{(i)}_{j+1/2}, \tag{5.6}$$

with

$$K^{(i)}_{j+1/2} = f(u^{(i)}_{j+1}) - f(u^{(i)}_j) \qquad \text{with} \quad u^{(1)}_j = u^n(x_j). \tag{5.7}$$

To evaluate the intermediate states, $u^{(i)}_j$, we exploit the fact that the reconstruction $u^n(x)$ is smooth except for jumps at the cell edges, $x_{j\pm 1/2}$. Thus, $u(x_j,t)$ remains smooth for $t \in [t^n, t^n+\Delta t]$, if Δt is small enough. As in all central schemes based on staggered grids, we can therefore evaluate the intermediate states $u^{(i)}_j$ integrating the conservation law (5.1) in its differential form, namely: $u_t = -f_x(u)$. Thus:

$$u^{(i)}_j = u^{(1)}_j + \Delta t \sum_{l=1}^{i-1} a_{i,l} \hat{K}^{(l)}_j, \qquad i = 2, \dots, \nu, \tag{5.8}$$

with the Runge-Kutta fluxes:

$$\hat{K}_j^{(l)} = -\left.\frac{\partial f(u^{(l)})}{\partial x}\right|_j \qquad \text{and} \qquad u_j^{(1)} = u^n(x_j) \quad l = 2, \ldots, \nu. \tag{5.9}$$

There are other ways to compute the Runge-Kutta fluxes for the computation of the intermediate values, other than the one exposed here. In particular, one could write

$$\frac{\partial f(u)}{\partial x} = A(u)\frac{\partial u}{\partial x}$$

and use the recontruction for the field u to compute an approximation of $\partial u/\partial x$. This approach avoids the reconstruction of the flux f, but requires the computation of the Jacobian matrix A.

At this stage, we have completed the time discretization of the scheme. Note that the scheme is in conservation form for the numerical solution, while the stage values are computed by a non-conservative scheme.

We now need to specify the space discretization, which is obtained with reconstruction techniques. The reconstruction must yield the following quantities:

- Starting from the cell averages $\{\bar{u}_j^n\}$, compute the staggered cell averages, $\{\bar{u}_{j+1/2}^n\}$, appearing in (5.6), as defined in (5.4).
- Starting from the cell averages $\{\bar{u}_j^n\}$, compute the point values, $\{u_j^n = u^n(x_j)\}$, which are needed to compute the start-up values $u_j^{(1)}$ in (5.8).
- Starting from the intermediate values $\{u_j^{(i)}\}$, compute the derivative of the fluxes $\{f_x(u^{(i)})|_j\}$. These quantities form the Runge-Kutta fluxes in (5.9).

Note that we can use different interpolation algorithms for each quantity we need to reconstruct. According to our numerical evidence (see also [52]) the key step is the reconstruction of $\{\bar{u}_{j+1/2}^n\}$.

The algorithm for CRK schemes can be conveniently written as a three-step method, as follows:

1. **Reconstruction step:** From the cell averages $\{\bar{u}_j^n\}$, compute the reconstruction $u^n(x)$. Use this to evaluate $u_j^{(1)} = u^n(x_j)$ and $\bar{u}_{j+1/2}^n$.
2. **Predictor step (stage values):** Compute the intermediate states $u_j^{(i)}$:

   For $i = 2, \ldots, \nu$:

   - compute $f(u_j^{(i)})\ \forall j$;
   - apply a suitable non-oscillatory interpolation to yield $\hat{K}_j^{(i)} = -f_x(u^{(i)})_j$;
   - compute $u_j^{(i)} = u_j^{(1)} + \Delta t \sum_{l=1}^{i-1} a_{i,l}\hat{K}_j^{(l)}$.

3. **Corrector step (numerical solution):** Assemble the fluxes $K^{(i)}_{j+1/2}$ defined above, and update the cell averages on the staggered grid:

$$\bar{u}^{n+1}_{j+1/2} = \bar{u}^n_{j+1/2} - \lambda \sum_{i=1}^{\nu} b_i K^{(i)}_{j+1/2}.$$

Note that the computation of $\hat{K}^{\nu}_j$ is not required. This improves the efficiency of the scheme, because it requires one less interpolation per time step, with respect to previous high-order Central WENO schemes [40]. We end this section by giving all details needed to code the CRK schemes we will test in the next sections.

5.3 A Second-Order Scheme: CRK2

The second-order scheme we propose has Nessyahu-Tadmor piecewise linear reconstruction:

$$u^n(x) = \sum_j \left(\bar{u}^n_j + u'_j(x - x_j)\right) \chi_{I_j}(x),$$

where u'_j is an approximate slope, computed for instance with the MinMod limiter. Time integration is given by Heun's scheme. Therefore the scheme is:

$$\begin{aligned}
\bar{u}^n_{j+1/2} &= \frac{1}{2}(\bar{u}^n_j + \bar{u}^n_{j+1}) + \frac{1}{8}(u'_j - u'_{j+1}), \\
u^{(1)}_j &= \bar{u}^n_j - \Delta t f_x(\bar{u}^n_j) \\
\bar{u}^{n+1}_{j+1/2} &= \bar{u}^n_{j+1/2} - \frac{\lambda}{2}(f(\bar{u}^n_{j+1}) + f(u^{(1)}_{j+1}) - f(\bar{u}^n_j) - f(u^{(1)}_j)),
\end{aligned}$$

where, for instance,

$$f_x(\bar{u}^n_j) = \frac{1}{h}\mathrm{MinMod}(f(\bar{u}^n_{j+1}) - f(\bar{u}^n_j), f(\bar{u}^n_j) - f(\bar{u}^n_{j-1})).$$

It is easy to prove that this scheme coincides with the Nessyahu-Tadmor scheme in the case of linear advection. Moreover, following the same steps appearing in [45], it is easy to prove that the CRK2 scheme is TVD, under a stricter CFL condition, provided that suitable limiters are used to compute both u'_j and $f_x(u)|_j$. Note that if the Modified Euler rule is used for time integration instead of the TVD Heun scheme, the CRK2 scheme coincides with Nessyahu-Tadmor scheme.

5.4 Higher-Order Schemes: CRK3, CRK4, CRK5

In this section, we describe a third-, fourth- and fifth-order scheme (respectively: CRK3, CRK4, and CRK5). All these schemes are built with Central WENO reconstructions. In the following, we review the main steps of the reconstruction, in order to allow the coding of the schemes we are proposing.

We start describing how $\bar{u}^n_{j+1/2}$ is computed. Then we will sketch how point values and flux derivatives are constructed.

In all three cases, the reconstruction from cell averages is a nonlinear convex combination of three interpolating polynomials. On the interval I_j:

$$u^n(x)|_{I_j} = R_j(x) = w_j^{-1} p_{j-1}(x) + w_j^0 p_j(x) + w_j^{+1} p_{j+1}(x), \qquad (5.10)$$

where the w_j^k are the nonlinear weights, which satisfy $\sum_k w_j^k = 1$.

For the third-order scheme, [41], p_{j-1} and p_{j+1} are two linear functions, while p_j is a parabola, namely:

$$\begin{aligned}
p_{j-1}(x) &= \bar{u}^n_j + \frac{1}{h}(\bar{u}^n_j - \bar{u}^n_{j-1})(x - x_j),\\
p_{j+1}(x) &= \bar{u}^n_j + \frac{1}{h}(\bar{u}^n_{j+1} - \bar{u}^n_j)(x - x_j),\\
p_j(x) &= \bar{u}^n_j - \frac{1}{12}(\bar{u}^n_{j+1} - 2\bar{u}^n_j + \bar{u}^n_{j-1}) + \frac{1}{2h}(\bar{u}^n_{j+1} - \bar{u}^n_{j-1})(x - x_j)\\
&\quad + \frac{1}{h^2}(\bar{u}^n_{j+1} - 2\bar{u}^n_j + \bar{u}^n_{j-1})(x - x_j)^2.
\end{aligned}$$

For the fourth- and fifth-order schemes, all polynomials $p_{j+k}(x), k = -1, 0, +1$ are parabolas. The coefficients of the parabola p_{j+k} are given by the following interpolation requirements:

$$\frac{1}{h}\int_{x_{j+k}+(l-\frac{1}{2})h}^{x_{j+k}+(l+\frac{1}{2})h} p_{j+k}(x)\, dx = \bar{u}_{j+k+l}, \qquad l = -1, 0, +1.$$

In all schemes considered here the reconstruction $u^n(x)$ is piecewise parabolic. Note that the reconstruction is conservative, since:

$$\frac{1}{h}\int_{x_{j+k+1/2}}^{x_{j+k-1/2}} R_j(x)\, dx = \bar{u}_{j+k}, \qquad k = -1, 0, +1.$$

The weights w_j^k are determined in order to:

- maximize accuracy in smooth regions
- prevent the onset of spurious oscillations.

Following [28], we define the weights with the following formulas:

$$w_j^k = \frac{\alpha_j^k}{\sum_{l=-1}^{1} \alpha_j^l} \qquad \Longrightarrow \qquad \sum_{l=-1}^{1} w_j^k = 1, \qquad (5.11)$$

where:

$$\alpha_j^k = \frac{C^k}{\left(\epsilon + \beta_j^k\right)^2}. \qquad (5.12)$$

Scheme	C^{-1}	C^{0}	C^{+1}	Accuracy
Reconstruction of $\bar{u}^n_{j+1/2}$				
CRK3	1/4	1/2	1/4	h^3
CRK4	3/16	5/8	3/16	h^5
CRK5	3/16	5/8	3/16	h^5
Reconstruction of $u^n(x_j)$				
CRK3	1/4	1/2	1/4	h^3
CRK4	3/16	5/8	3/16	h^4
CRK5	-9/80	49/40	-9/80	h^5
Reconstruction of $f_x(u(x,.)\|_{x_j}$				
CRK3	1/4	1/2	1/4	h^2
CRK4	1/6	1/3	1/6	h^4
CRK5	1/6	1/3	1/6	h^4

Table 5.1: Accuracy constants.

The C^k's are called accuracy constants. They are determined in order to maximize accuracy in smooth regions, and they depend on the particular quantity being reconstructed, see Table 5.1. More details will be given below.

The parameter ϵ prevents a vanishing denominator. It is $\epsilon = 10^{-4}$ for the third-order scheme, while $\epsilon = 10^{-6}$ for the fourth- and fifth-order schemes.

Finally, $\beta_j^k, k = -1, 0, 1$ are the smoothness indicators. β_j^k is a measure of the regularity of the polynomial p_{j+k} on the interval I_j. It is defined (see Section 4.3 and reference [28]), as a rescaled measure of the H^2 seminorm of the polynomial on I_j:

$$\beta_j^k = \sum_{l=1}^{2} \int_{x_{j-1/2}}^{x_{j+1/2}} h^{2l-1} \left(\frac{d^l p_{k+j}}{dx^l} \right)^2 dx, \qquad k = -1, 0, 1. \tag{5.13}$$

Thus $\beta_j^k = O(h^2)$ on smooth regions, while $\beta_j^k = O(1)$ if the data in the stencil of p_{k+j} contain a jump discontinuity. As a consequence of the normalization factor, $w_j^k = C^k + O(h^s)$, if the stencil S_{j+k} on which p_{j+k} is defined contains smooth data ($s = 1$ for the third-order reconstruction, while $s = 2$ in the fourth- and fifth-order reconstructions).

The accuracy constants C^k are the solution of an interpolation problem. Suppose $u(x)$ is a given smooth function, with cell averages $\{\bar{u}_j\}$. If the reconstruction is computed to evaluate the staggered cell average $\bar{u}_{j+1/2}$, then we require:

$$\frac{1}{h}\int_{x_j}^{x_{j+1/2}} R_j(x)\,dx = \frac{1}{h}\int_{x_j}^{x_{j+1/2}} u(x)\,dx + O(h^s),$$

where $s = 3$, for the third-order scheme, while $s = 5$ for CRK4 and CRK5.

For the interpolation of point values, we use a different set of constants. In this case, the constants are determined to yield:

$$R_j(x_j) = u(x_j) + O(h^s),$$

where $s = 3, s = 4$ and $s = 5$ for CRK3, CRK4 and CRK5, respectively. For the third- and fourth-order scheme, any symmetric combination of constants, adding up to 1 will do. For simplicity, we use the same constants dictated by the evaluation of $\bar{u}^n_{j+1/2}$, as listed in Table 5.1.

Finally, for flux derivatives, the polynomials p_{j+k} are determined by the following interpolation requirements:

$$p_{j+k}(x_{j+k+l}) = f(u(x_{j+k+l})), \qquad l = -1, 0, 1,$$

for the fourth- and fifth-order schemes. For the third-order scheme the interpolation requirements are:

$$\begin{aligned} &p_{j+1}(x_{j+l}) = f(u(x_{j+l})), \quad p_{j-1}(x_{j-l}) = f(u(x_{j-l})), \quad l = 0, 1, \\ &p_j(x_{j+l}) = f(u(x_{j+l})), \qquad l = -1, 0, 1. \end{aligned}$$

The accuracy constants for the reconstruction of flux derivatives satisfy:

$$\partial_x R_j\big|_{x_j} = \partial_x f(u(x))\big|_{x_j} + O(h^s),$$

where $s = 4$ for the fourth- and fifth-order schemes, while $s = 2$ for the third-order schemes. The resulting constants are given in Table 5.1.

Particular care is needed in the reconstruction of point values for the fifth-order CRK5 scheme. In this case in fact two accuracy constants for the reconstruction of point values are negative (see Table 5.1), and a straightforward application of the algorithm with the nonlinear weights yields a scheme that can become unstable, especially in the presence of interactions between discontinuities.

To avoid this problem, Shu and co-workers proposed a splitting into positive and negative weights. We refer to [60] and [50] for more details.

The time integration is carried out with the Runge-Kutta schemes listed in Table 5.2. RK3 is the third-order TVD Runge-Kutta scheme appearing in [59]. RK4 is the standard fourth-order Runge-Kutta scheme.

Linear stability analysis can be performed to establish the maximum CFL number allowed for each scheme. This information will be then used for the computation of the time step. The results of the analysis for the various schemes are

Coefficients of Runge-Kutta schemes

Scheme	Lower triangular part of A	b
RK2	1	1/2 1/2
RK3	1 1/4 1/4	1/6 1/6 2/3
RK4	1/2 0 1/2 0 0 1	1/6 1/3 1/3 1/6
RK5	1/2 3/16 1/16 0 0 1/2 0 −3/16 6/16 9/16 1/7 4/7 6/7 −12/7 8/7	$\frac{7}{90}$ 0 $\frac{32}{90}$ $\frac{12}{90}$ $\frac{32}{90}$ $\frac{7}{90}$

Table 5.2: Coefficients of Runge-Kutta schemes.

the following. Let $\rho(\xi)$ be the amplification factor of the scheme applied ti the linear advection equation. Then, for $\lambda \leq \lambda_0$, $\rho(\xi) \leq 1$ for all wave numbers ξ:

$$\begin{aligned}
\text{CRK2}: \quad \lambda_0 &= \frac{1}{2} = 0.5 \\
\text{CRK3}: \quad \lambda_0 &= \frac{3}{7} \simeq 0.42 \\
\text{CRK4}: \quad \lambda_0 &= \frac{12}{25} = 0.48 \\
\text{CRK5}: \quad \lambda_0 &= \frac{60}{149} \simeq 0.40.
\end{aligned} \tag{5.14}$$

5.5 Numerical Tests

All CRK schemes perform well with Burgers' equation, providing sharp shock resolution. Central schemes are somehow less effective than upwind schemes in treating linear discontinuities, because of their intrinsic dissipation. Nevertheless, reasonably good results may be obtained even for such problems, as is illustrated in the following test on contact discontinuities.

In the test chosen, we consider the linear advection equation $u_t + u_x = 0$, with initial condition:

$$u(x, t = 0) = \begin{cases} \cos(\frac{\pi}{2}x) & -1 \leq x \leq 0, \\ \sin(\pi x) & 0 < x \leq 1, \end{cases}$$

on $[-1, 1]$ with periodic boundary conditions. The solution restricted to $[-1, 1]$ contains a contact discontinuity and an angular point, induced by the boundary conditions. The results are shown in Figure 5.2 at $T = 5.5$, i.e., after quite a long integration time, for $\lambda = 0.9\lambda_0$, with λ_0 given in (5.14).

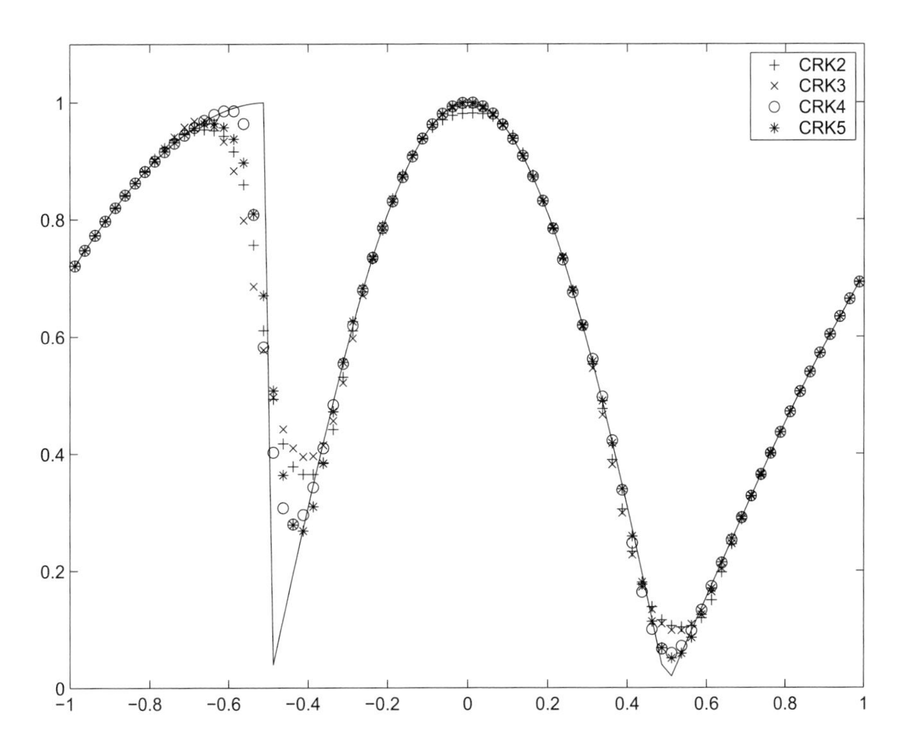

Figure 5.2: Linear advection equation $\lambda = 0.9\lambda_0$. Solution given by CRK2, CRK3, CRK4, and CRK5 for $N = 80$.

5.6 Systems of Equations

CRK schemes can be applied to systems of equations. In the simplest form, the reconstruction can be performed componentwise, on the conservative variables. As we shall see, less oscillatory results are obtained by using characteristic variables in the reconstruction.

The one-dimensional Euler equations for an ideal gas consist of three conservation laws. The vector of unknowns is $u = (\rho, \rho v, E)^T$, where ρ is the density, v is the velocity and E is the total energy per unit volume, given by $E = \frac{1}{2}\rho v^2 + \rho e$, with e the internal energy, linked to the pressure p by the equation of state $p = p(\rho, e)$. For a polytropic gas $p = \rho e(\gamma - 1)$, with $\gamma = c_p/c_v$. The value $\gamma = 7/5$, valid for a diatomic gas such as air, has been used in the numerical tests. The flux is $f(u) = (\rho v, \rho v^2 + p, v(E + p))$.

5.7 Componentwise Application

The simplest approach to the integration of systems of conservation laws with CRK schemes is to apply the schemes component by component to each equation of system (5.1). A Global Smoothness Indicator can be obtained [40] as

$$\beta_k^j = \frac{1}{m_c} \sum_{r=1}^{m_c} \frac{1}{||\bar{u}_r||_2^2} \left(\sum_{l=1}^{2} \int_{I_j} h^{2l-1} \left(\frac{d^l p_{j+k,r}}{dx^l} \right)^2 dx \right) \tag{5.15}$$

for $k = -1, 0, 1$. Here r denotes the r-th component of the solution and of the vector-valued interpolating polynomial, and m_c denotes the number of equations of the hyperbolic system. Comparing with (5.13), we see that the Global Smoothness Indicator is just a weighted average of the Smoothness Indicators given by each component. The weights in the average are chosen in order to obtain a dimensionless quantity. This ensures that the indicators are invariant with respect to units of measure.

The first test we show is by Shu and Osher [59]. It describes the interaction of an acoustic wave with a shock. The solution has a rich structure, which is better resolved by a high-accuracy scheme. The initial condition is $u = u_L$ for $x \leq 0.1$, and $u = u_R$ for $x > 0.1$. The computational domain is $[0, 1]$, with free-flow boundary conditions. The left (L) and right (R) states are given by:

$$\begin{pmatrix} \rho \\ v \\ p \end{pmatrix}_L = \begin{pmatrix} 3.857143 \\ 2.629369 \\ 10.3333 \end{pmatrix}, \quad \begin{pmatrix} \rho \\ v \\ p \end{pmatrix}_R = \begin{pmatrix} 1 + 0.2\sin(50x) \\ 0 \\ 1 \end{pmatrix}.$$

The reciprocal of the maximum characteristic speed for this flow is $c \simeq 0.219$. Thus we used the mesh ratio $\lambda = 0.2\lambda_0$, where the appropriate value of λ_0 can be found, as usual, in (5.14). The solution is plotted at $T = 0.18$. The results are shown in Figure 5.3. The solid line is the reference solution, obtained with CRK4 and 1600 grid points. The dotted line is the numerical solution, computed using 400 grid points.

The figure clearly shows that there is a noticeable increase in resolution, using high-order schemes. Although the difference between the numerical solutions obtained with CRK2 and CRK3 is very small, there is a very strong improvement with the fourth-order scheme. In fact, the structure behind the strong shock is not

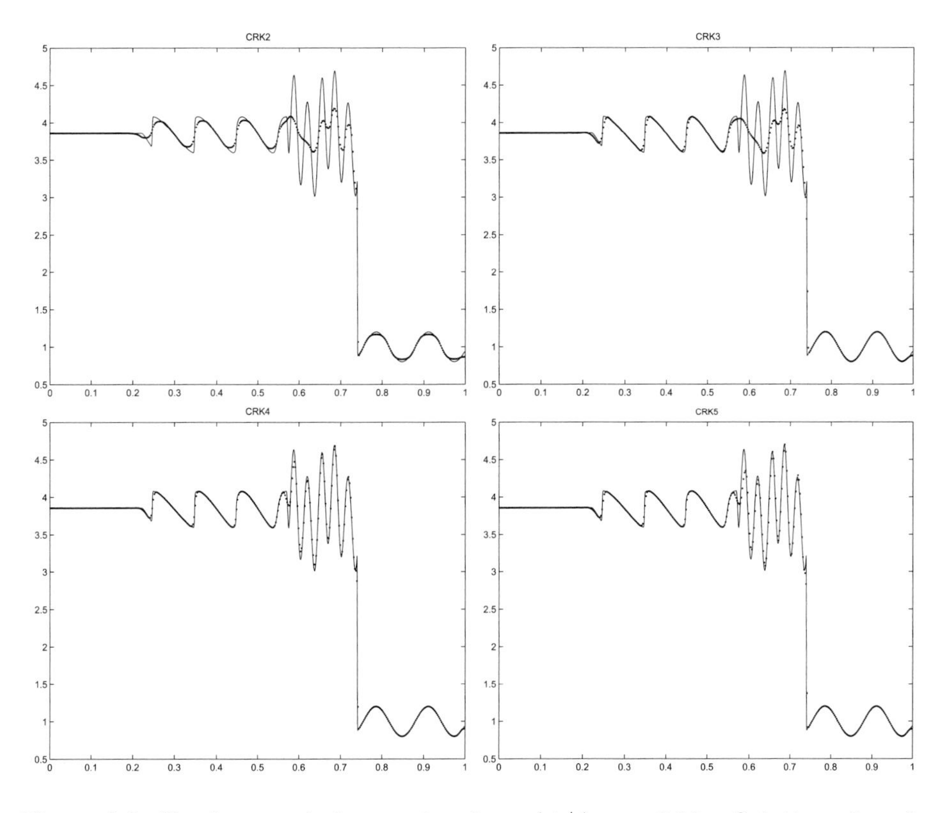

Figure 5.3: Shock-acoustic interaction $\lambda \equiv \Delta t/\Delta x = 0.2\lambda_0$. Solution given by CRK2, CRK3, CRK4, and CRK5 for $N = 400$.

resolved by the low-order schemes. Instead, its complexity is well represented by the fourth- and the fifth-order scheme, with no need of a very fine mesh.

Next, we consider a Riemann problem due to Lax [35]. Here the initial condition is $u = u_L$ for $x \leq 0.5$, and $u = u_R$ for $x > 0.5$. The computational domain is $[0, 1]$, with free-flow boundary conditions. The left (L) and right (R) states are given by:

$$u_L = \begin{pmatrix} 0.445 \\ 0.311 \\ 8.928 \end{pmatrix}, \qquad u_R = \begin{pmatrix} 0.5 \\ 0. \\ 1.4275 \end{pmatrix}.$$

The inverse of the maximum eigenvalue for this flow is approximately $c = 0.21$. As mesh ratio, we thus pick $\lambda = 0.2\lambda_0$, where λ_0 is, as usual, the mesh ratio obtained with linear stability analysis, see (5.14). This means that the Courant number is very close to the critical one.

The results are shown in Figure 5.4, where a detail in the density peak is shown. The whole solution can be seen for instance in [40] or [52]. The detail we

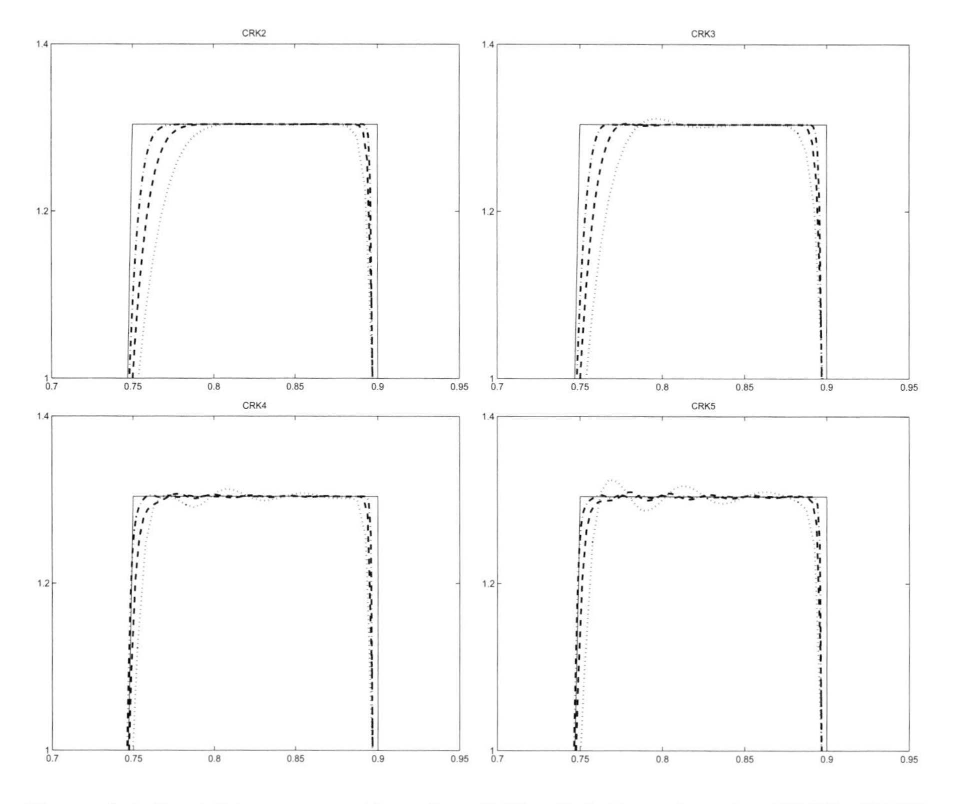

Figure 5.4: Lax' Riemann problem $\lambda = 0.2\lambda_0$. Solution given by CRK2, CRK3, CRK4, and CRK5 for $N = 200$ (dotted line), $N = 400$ (dashed line) and $N = 800$ (dash-dotted line).

show is a zoom on the region containing spurious oscillations The discontinuities appearing on the left and on the right of the density peak are respectively a contact and a shock wave. The solution is shown, for all schemes studied here, for $N = 200$ (dotted line), $N = 400$ (dashed line) and $N = 800$ (dash-dotted line). The high-order schemes clearly exhibit small amplitude spurious oscillations in this test problem. These oscillations are clearly of ENO type, in the sense that their amplitude decreases as the grid is refined. It can also be noted that the shock is, as expected, better resolved than the contact wave, and the resolution of the contact improves with the order of accuracy.

5.8 Projection Along Characteristic Directions

While the results shown in Figure 5.3 are quite satisfactory, the numerical solutions shown at the bottom of Figure 5.4 can be improved. The componentwise application of CRK schemes seems to be inadequate whenever discontinuities are separated by regions of almost constant states. The result will be improved by using characteristic variables in order to perform the reconstruction. It can be shown that it is enough to use characteristic variables for the computation of the staggered cell average $\bar{u}^n_{j+1/2}$ from cell averages, because this is where the upwind information of the characteristic variables is mostly needed.

The interested reader may consult, for example, [50] or [51] for an example of efficient implementation of the characteristic reconstruction.

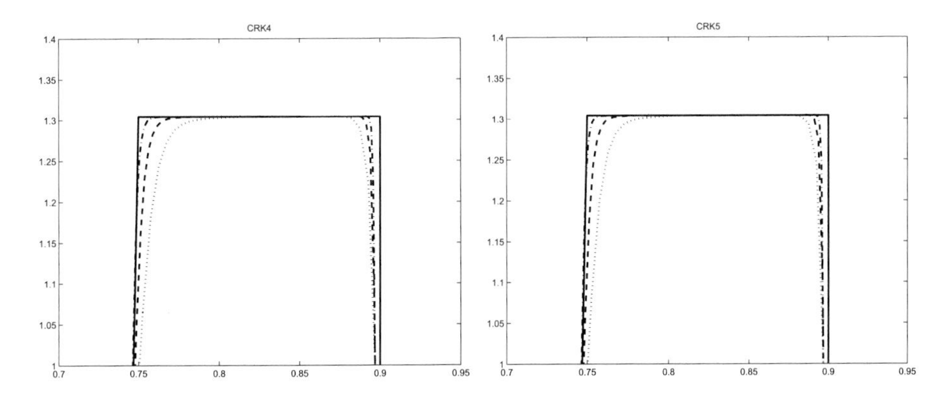

Figure 5.5: Lax' Riemann problem $\lambda = 0.2\lambda_0$. Solution given by CRK4 (left) and and CRK5 (right) for $N = 200$ (dotted line), $N = 400$ (dashed line) and $N = 800$ (dash-dotted line). The interpolation is computed with projection along characteristic directions.

Chapter 6

Systems with Stiff Source

The development of efficient numerical schemes for such systems is challenging, since in many applications the relaxation time varies from values of order 1 to very small values if compared to the time scale determined by the characteristic speeds of the system. In this second case the hyperbolic system with relaxation is said to be stiff, and typically its solutions are well approximated by solutions of a suitably reduced set of conservation laws called *equilibrium system* [13].

Usually it is extremely difficult, if not impossible, to split the problem into separate regimes and to use different solvers in the stiff and non-stiff regions. Thus one has to use the original relaxation system in the whole computational domain.

Splitting methods have been widely used for such problems. They are attractive because of their simplicity and robustness. Strang splitting provides second-order accuracy if each step is at least second-order accurate [62]. This property is maintained under fairly mild assumptions even for stiff problems [27]. However, Strang splitting applied to hyperbolic systems with relaxation reduces to first-order accuracy when the problem becomes stiff. The reason is that the kernel of the relaxation operator is non-trivial, which corresponds to a singular matrix in the linear case, and therefore the assumptions in [27] are not satisfied.

Furthermore with a splitting strategy it is difficult to obtain higher-order accuracy even in non-stiff regimes (high-order splitting schemes can be constructed, see [17], but they are seldom used because of stability problems).

Recently developed Runge-Kutta schemes overcome these difficulties, providing basically the same advantages of the splitting schemes, without the drawback of the order restriction [12, 29, 65].

A general methodology that can be used for the treatment of hyperbolic systems of balance laws with stiff source is obtained by making use of Implicit-Explicit (IMEX) Runge-Kutta schemes. The hyperbolic part, which is in general non-stiff[1] can be treated by the explicit part of the scheme, while the stiff source is treated by the implicit part.

In order to guarantee TVD property in time for the hyperbolic part, total variation diminishing Runge-Kutta schemes (also called Strongly Stability Preserving (SSP) schemes) have been developed (see Section 4.6 and [22, 23, 61]). In this section we shall mainly restrict to the case of hyperbolic systems with stiff relaxation. For such systems, suitable schemes are obtained by coupling SSP schemes for the hyperbolic part, with L-stable schemes for the relaxation.

Some applications to systems with stiff relaxation will be presented.

6.1 Systems of Balance Laws

Many physical systems are described by a system of balance laws of the form

$$\frac{\partial u}{\partial t} + \frac{\partial f(u)}{\partial x} = \frac{1}{\epsilon} g(u), \tag{6.1}$$

where we have written explicitly a factor $1/\epsilon$ in front of the source term, to express that such a source may be stiff. Hyperbolic systems with a source can be treated by both finite-volume and finite-difference methods.

Finite-volume. Integrating Equation (6.1) in space in cell I_j and dividing by Δx one has

$$\frac{d\bar{u}_j}{dt} = -\frac{F_{j+1/2} - F_{j-1/2}}{h} + \frac{1}{\epsilon}\overline{g(u)_j}, \tag{6.2}$$

where $\overline{g(u)_j}$ denotes the j-th cell average of the source.

In order to convert this expression into a numerical scheme, one has to approximate the right-hand side with a function of the cell averages $\{\bar{u}(t)\}_j$, which are the basic unknowns of the problem.

The right-hand side of Equation (6.2) contains the average of the source term $\overline{g(u)}$ instead of the source term evaluated at the average of u, $g(\bar{u})$. The two quantities agree within second-order accuracy

$$\overline{g(u)}_j = g(\bar{u}_j) + O(\Delta x^2).$$

This approximation can be used to construct schemes up to second order.

First-order (in space) semidiscrete schemes can be obtained using the numerical flux function $F(\bar{u}_j, \bar{u}_{j+1})$ in place of $f(u(x_{j+1/2}, t))$,

$$\frac{d\bar{u}_j}{dt} = -\frac{F(\bar{u}_j, \bar{u}_{j+1}) - F(\bar{u}_{j-1}, \bar{u}_j)}{\Delta x} + \frac{1}{\varepsilon} g(\bar{u}_j). \tag{6.3}$$

[1] If one is interested in resolving all the waves or if the wave speeds are of the same order of magnitude, then the hyperbolic part is not stiff, in the sense that the CFL stability condition is in agreement with accuracy requirements. If fast waves carry a negligible signal, then the problem becomes stiff, since the CFL restriction on the time step is due to the fast waves, while accuracy requires to resolve much slower time scales, typical of the evolution of the slow waves. We shall not treat such a case here.

Second-order schemes are obtained by using a piecewise linear reconstruction in each cell, and evaluating the numerical flux on the two sides of the interface:

$$\frac{d\bar{u}}{dt} = -\frac{F(u^-_{j+1/2}, u^+_{j+1/2}) - F(u^-_{j-1/2}, u^+_{j-1/2})}{\Delta x} + \frac{1}{\varepsilon} g(\bar{u}_j).$$

The quantities at cell edges are computed by piecewise linear reconstruction. For example,

$$u^-_{j+1/2} = \bar{u}_j + \frac{\Delta x}{2} u'_j,$$

where the slope u'_j is a first-order approximation of the space derivative of $u(x,t)$, and can be computed by suitable slope limiters (see, for example, [38] for a discussion on TVD slope limiters.)

For schemes of order higher than second, a suitable quadrature formula is required to approximate $\overline{g(u)}_j$. For example, for third- and fourth-order schemes, one can use Simpson's rule

$$\overline{g(u)}_j \approx \frac{1}{6}(g(u^+_{j-1/2}) + 4g(u_j) + g(u^-_{j+1/2})),$$

where the pointwise values $u^+_{j-1/2}$, u_j, $u^-_{j+1/2}$ are obtained from the reconstruction.

For a general problem, this has the effect that the source term couples the cell averages of different cells, thus making almost impractical the use of finite-volume methods for high-order schemes applied to stiff sources, where the source is treated implicitly.

Note, however, that in many relevant cases of hyperbolic systems with relaxation the implicit step, thanks to the conservation properties of the system, can be explicitly solved, and finite-volume methods can be successfully used, even in the high-order case. We mention here all relaxation approximation of Jin-Xin type [30], some simple discrete velocity models, such as Carlemann and Broadwell models [20, 10], monatomic gas in Extended Thermodynamics [44], semiconductor models [1, 2], and shallow water equations [29].

Finite-difference. A finite-difference method applied to (6.1) becomes

$$\frac{du_j}{dt} = -\frac{\hat{F}_{j+1/2} - \hat{F}_{j-1/2}}{h} + \frac{1}{\epsilon} g(u_j), \tag{6.4}$$

where the source term is computed pointwise and no strange integration is needed. Both systems (6.2) and (6.4) consist of two terms on the right-hand side. They have the structure of a system of $J \times m_c$ ordinary differential equations. If the stability restriction the time step due to the presence of these two terms are comparable (when using an explicit scheme), then the source is not stiff, and one can use a standard explicit ODE solver (for example a TVD Runge-Kutta scheme). If, on the other hand, the stability restriction on Δt imposed by the presence of $g(u_j)/\epsilon$

are much more severe (for example because $g(u_j)/\epsilon$ represents a relaxation with a small relaxation time), then it is better to use a time discretization which is explicit in the hyperbolic part and implicit in the source term (IMEX).

Here we use a method of lines approach, and we write system (6.4) in vector form as

$$\frac{dU}{dt} = H(U) + \frac{1}{\epsilon}G(U) \tag{6.5}$$

where $U \in \mathbb{R}^{Jm_c}$, J being the number of space cells, and m_c being the number of components of the function $u(x,t)$. For finite-difference schemes, the function $G(U)$ is diagonal ($m_c = 1$) or block diagonal ($m_c > 1$), and therefore the implicit step involving only function G requires the solution of a relatively small ($m_c \times m_c$) system.

6.2 IMEX Runge-Kutta Schemes

An IMEX Runge-Kutta scheme consists of applying an implicit discretization to the source terms and an explicit one to the non-stiff term. When applied to system (6.5) it takes the form

$$U^{(i)} = U^n + \Delta t \sum_{j=1}^{i-1} \tilde{a}_{ij} H(U^{(j)}) + \Delta t \sum_{j=1}^{i} a_{ij} \frac{1}{\varepsilon} G(U^{(j)}), \tag{6.6}$$

$$U^{n+1} = U^n + \Delta t \sum_{i=1}^{\nu} \tilde{b}_i H(U^{(i)}) + \Delta t \sum_{i=1}^{\nu} b_i \frac{1}{\varepsilon} G(U^{(i)}). \tag{6.7}$$

The matrices $\tilde{A} = (\tilde{a}_{ij})$, $\tilde{a}_{ij} = 0$ for $j \geq i$ and $A = (a_{ij})$ are $\nu \times \nu$ matrices such that the resulting scheme is explicit in H, and implicit in G. An IMEX Runge-Kutta scheme is characterized by these two matrices and the coefficient vectors $\tilde{b} = (\tilde{b}_1, \ldots, \tilde{b}_\nu)^T$, $b = (b_1, \ldots, b_\nu)^T$.

The implicit scheme is diagonally implicit (DIRK), i.e., $a_{ij} = 0$, for $j > i$ [24]. This will simplify the solution of the algebraic equations associated to the implicit step, and it will guarantee that the function H will appear explicitly in the scheme (a fully implicit scheme will couple the evaluation of the function H at different stages.)

IMEX Runge-Kutta schemes can be represented by a double *tableau* in the usual Butcher notation,

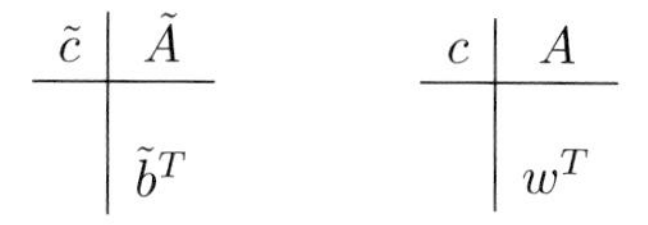

$$\begin{array}{c|c} \tilde{c} & \tilde{A} \\ \hline & \tilde{b}^T \end{array} \qquad \begin{array}{c|c} c & A \\ \hline & w^T \end{array}$$

where the coefficients $\tilde{c}$ and c used for the treatment of non-autonomous systems are given by the usual relation

$$\tilde{c}_i = \sum_{j=1}^{i-1} \tilde{a}_{ij}, \quad c_i = \sum_{j=1}^{i} a_{ij}. \tag{6.8}$$

Order Conditions

The general technique to derive order conditions for Runge-Kutta schemes is based on the Taylor expansion of the exact and numerical solution.

In particular, conditions for schemes of order p are obtained by imposing that the solution of system (6.1) at time $t = t_0 + \Delta t$, with a given initial condition at time t_0, agrees with the numerical solution obtained by one step of a Runge-Kutta scheme with the same initial condition, up to order Δt^p.

Here we report the order conditions for IMEX Runge-Kutta schemes up to order $p = 3$, which is already considered high-order for PDE problems.

We apply scheme (6.6)–(6.7) to system (6.1), with $\varepsilon = 1$. We assume that the coefficients $\tilde{c}_i$, c_i, $\tilde{a}_{ij}$, a_{ij} satisfy conditions (6.8). Then the order conditions are the following.

First-order.

$$\sum_{i=1}^{\nu} \tilde{b}_i = 1, \quad \sum_{i=1}^{\nu} b_i = 1. \tag{6.9}$$

Second-order.

$$\sum_i \tilde{b}_i \tilde{c}_i = 1/2, \quad \sum_i b_i c_i = 1/2, \tag{6.10}$$

$$\sum_i \tilde{b}_i c_i = 1/2, \quad \sum_i b_i \tilde{c}_i = 1/2. \tag{6.11}$$

Third-order.

$$\sum_{ij} \tilde{b}_i \tilde{a}_{ij} \tilde{c}_j = 1/6, \quad \sum_i \tilde{b}_i \tilde{c}_i \tilde{c}_i = 1/3, \quad \sum_{ij} b_i a_{ij} c_j = 1/6, \quad \sum_i b_i c_i c_i = 1/3, \tag{6.12}$$

$$\begin{aligned}
&\sum_{ij} \tilde{b}_i \tilde{a}_{ij} c_j = 1/6, \quad \sum_{ij} \tilde{b}_i a_{ij} \tilde{c}_j = 1/6, \quad \sum_{ij} \tilde{b}_i a_{ij} c_j = 1/6,\\
&\sum_{ij} b_i \tilde{a}_{ij} c_j = 1/6, \quad \sum_{ij} b_i a_{ij} \tilde{c}_j = 1/6, \quad \sum_{ij} b_i \tilde{a}_{ij} \tilde{c}_j = 1/6,\\
&\sum_i \tilde{b}_i c_i c_i = 1/3, \quad \sum_i \tilde{b}_i \tilde{c}_i c_i = 1/3, \quad \sum_i b_i \tilde{c}_i \tilde{c}_i = 1/3,\\
&\sum_i b_i \tilde{c}_i c_i = 1/3.
\end{aligned} \tag{6.13}$$

IMEX-RK	Number of coupling conditions			
order	General case	$\tilde{b}_i = b_i$	$\tilde{c} = c$	$\tilde{c} = c$ and $\tilde{b}_i = b_i$
1	0	0	0	0
2	2	0	0	0
3	12	3	2	0
4	56	21	12	2
5	252	110	54	15
6	1128	528	218	78

Table 6.1: Number of coupling conditions in IMEX Runge-Kutta schemes.

Conditions (6.9), (6.10), (6.12) are the standard order conditions for the two *tableau*, each of them taken separately. Conditions (6.11) and (6.13) are new conditions that arise because of the coupling of the two schemes.

The order conditions will simplify a lot if $\tilde{c} = c$. For this reason only such schemes are considered in [5]. In particular, we observe that, if the two tableau differ only for the value of the matrices A, i.e., if $\tilde{c}_i = c_i$ and $\tilde{b}_i = b_i$, then the standard order conditions for the two schemes are enough to ensure that the combined scheme is third order. Note, however, that this is true only for schemes up to third order.

Higher-order conditions can be derived as well using a generalization of Butcher 1-trees to 2-trees, see [31]. However the number of coupling conditions increase dramatically with the order of the schemes. The relation between coupling conditions and accuracy of the schemes is reported in Table 6.1.

6.3 Hyperbolic Systems with Relaxation

In this section we give sufficient conditions for asymptotic preserving and asymptotic accuracy properties of IMEX schemes. This properties are strongly related to L-stability of the implicit part of the scheme.

6.3.1 Zero Relaxation Limit

Consider a simple 2×2 system:

$$\begin{aligned} v_t + w_x &= 0 \\ w_t + v_x &= (bv - w)/\epsilon. \end{aligned} \tag{6.14}$$

Formally, as $\epsilon \to 0$, the second equation implies $w = bv$. Substituting this relation in the first equation one obtains a closed equation for v,

$$v_t + bv_x = 0, \tag{6.15}$$

In order to show that the solution to Equation (6.15) is the limit of the solution of system (6.14), one has to impose that $|b| < 1$, which corresponds to imposing the so-called *subcharacteristic condition.*

This condition can be easily understood using the following argument. Let us consider Equation (6.14), and let us look for a solution in terms of Fourier modes:

$$v(x,t) = \hat{v}(t)\exp(\iota kx), \quad w(x,t) = \hat{w}(t)\exp(\iota kx).$$

The equations for the Fourier modes are

$$\begin{aligned}\hat{v}_t &= -\iota k\hat{w},\\ \hat{w}_t &= -\iota k\hat{v} + \frac{1}{\epsilon}(b\hat{v} - \hat{w}).\end{aligned}$$

Introducing the vector $U = (\hat{v}, \hat{w})^T$, the system can be written in the form

$$\frac{dU}{dt} = AU, \quad \text{with } A = \begin{pmatrix} 0 & -\iota k \\ -\iota k + b/\epsilon & -1/\epsilon \end{pmatrix}.$$

Such a system will have unstable modes if there is an eigenvalue λ with positive real part. The characteristic equation for the matrix A takes the form

$$\lambda + \iota kb + \epsilon(\lambda^2 + k^2) = 0.$$

If $\epsilon = 0$, one has $\lambda = -\iota kb$, corresponding to undamped oscillations. For small values of ϵ one can write

$$\begin{aligned}\lambda &= -\iota kb - \epsilon(\lambda^2 + k^2)\\ &= -\iota kb - \epsilon(1 - b^2)k^2 + O(\epsilon^2),\end{aligned}$$

therefore, neglecting higher-order terms, the real part of λ is positive if $|b| > 1$. A similar expansion in ϵ can be performed directly at the level of the system (6.14), by using the so called *Chapman-Enskog expansion* [11]. In the case of Equation (6.14), the Chapman-Enskog expansion can be easily obtained by observing that, from the second equation, one has $w = bv + O(\epsilon)$. More precisely, using this observation and the equations (6.14), one has

$$\begin{aligned}w &= bv - \epsilon(w_t + v_x)\\ &= bv - \epsilon(bv_t + v_x) + O(\epsilon^2)\\ &= bv - \epsilon(-bw_x + v_x) + O(\epsilon^2)\\ &= bv - \epsilon(1 - b^2)v_x + O(\epsilon^2).\end{aligned}$$

Substituting this last relation in the first equation, one has

$$v_t + bv_x = \epsilon(1-b^2)v_{xx} + O(\epsilon^2).$$

Neglecting terms of higher order in ϵ one obtains a convection-diffusion equation with viscosity coefficient $\nu = \epsilon(1-b^2)$. Well-posedness of the initial value problem requires that $\nu \geq 0$, i.e., $|b| \leq 1$.

More generally, let us consider here one-dimensional hyperbolic systems with relaxation of the form (6.1). The operator $g : \mathbb{R}^N \to \mathbb{R}^N$ is called a relaxation operator, and consequently (6.1) defines a relaxation system, if there exists a constant $n \times m$ matrix Q with $\operatorname{rank}(Q) = n < m$ such that

$$Qg(u) = 0 \quad \forall\, u \in \mathbb{R}^m. \tag{6.16}$$

This gives n independent conserved quantities $u = Qu$. Moreover we assume that equation $g(u) = 0$ can be uniquely solved in terms of $\tilde{u}$, i.e.,

$$u = \mathcal{E}(\tilde{u}) \quad \text{such that} \quad g(\mathcal{E}(\tilde{u})) = 0. \tag{6.17}$$

The image of $\mathcal{E}$ represents the manifold of local equilibria of the relaxation operator g.

Using (6.16) in (6.1) we obtain a system of n conservation laws which is satisfied by every solution of (6.1),

$$\partial_t(Qu) + \partial_x(Qf(u)) = 0. \tag{6.18}$$

For vanishingly small values of the relaxation parameter ε from (6.1) we get $g(u) = 0$ which by (6.17) implies $u = \mathcal{E}(\tilde{u})$. In this case system (6.1) is well approximated by the equilibrium system [13]

$$\partial_t \tilde{u} + \partial_x \tilde{f}(\tilde{u}) = 0, \tag{6.19}$$

where $\tilde{f}(\tilde{u}) = Qf(\mathcal{E}(\tilde{u}))$.

System (6.19) is the formal limit of system (6.4) as $\varepsilon \to 0$. The solution $u(x,t)$ of the system will be the limit of Qu, with u solution of system (6.4), provided a suitable condition on the characteristic velocities of systems (6.4) and (6.19) is satisfied (the so-called *subcharacteristic condition*, see [64, 13].)

6.3.2 Asymptotic Properties of IMEX Schemes

We start with the following

Definition. We say that an IMEX scheme for system (6.1) in the form (6.6)–(6.7) is *asymptotic preserving* (AP) if in the limit $\varepsilon \to 0$ the scheme becomes a consistent discretization of the limit equation (6.19).

Note that this definition does not imply that the scheme preserves the order of accuracy in t in the stiff limit $\epsilon \to 0$. In the latter case the scheme is said *asymptotically accurate.*

In order to give sufficient conditions for the AP and asymptotically accurate property, we make use of the following simple

Lemma 6.1. *If all diagonal elements of the triangular coefficient matrix A that characterize the* DIRK *scheme are nonzero, then*

$$\lim_{\epsilon \to 0} g(u^{(i)}) = 0. \tag{6.20}$$

Proof. In the limit $\epsilon \to 0$ from (6.6) we have

$$\sum_{j=1}^{i} a_{ij} g(u^j) = 0, \quad i = 1, \dots, \nu.$$

Since the matrix A is non-singular, this implies $g(u^i) = 0$, $i = 1, \dots, \nu$. □

In order to apply the previous lemma, the vectors of c and $\tilde{c}$ cannot be equal. In fact $\tilde{c}_1 = 0$ whereas $c_1 \neq 0$. Note that if $c_1 = 0$ but $a_{ii} \neq 0$ for $i > 1$, then we still have $\lim_{\epsilon \to 0} g(u^{(i)}) = 0$ for $i > 1$ but $\lim_{\epsilon \to 0} g(u^{(1)}) \neq 0$ in general. The corresponding scheme may be inaccurate if the initial condition is not "well prepared" ($g(u_0) \neq 0$). In this case the scheme is not able to treat the so-called "initial layer" problem, and degradation of accuracy in the stiff limit is expected (see, for example, [12, 47, 46].) On the other hand, if the initial condition is "well prepared" ($g(u^{(0)}) = 0$), then relation (6.20), $i = 1, \dots, \nu$, holds even if $a_{11} = c_1 = 0$. In practice, if one uses a scheme that does not handle well an initial layer, then one should take small time steps at the beginning of the computation, during a short transient time. The automatic reduction of the time step during the transient should be taken into account by a good time-step control for ODE solvers (see, for example, in [24]).

Next we can state the following

Theorem 6.2. *If* $\det A \neq 0$ *in the limit* $\epsilon \to 0$, *the* IMEX *scheme* (6.6)–(6.7) *applied to system* (6.1) *becomes the explicit* RK *scheme characterized by* $(\tilde{A}, \tilde{w}, \tilde{c})$ *applied to the limit equation* (6.19).

For the proof see [49]. Clearly one may claim that if the implicit part of the IMEX scheme is A-stable or L-stable, the previous theorem is satisfied. Note however that this is true only if the *tableau* of the implicit integrator does not contain any column of zeros that makes it reducible to a simpler A-stable or L-stable form. Some remarks are in order.

Remarks.

i) There is a close analogy between hyperbolic systems with stiff relaxation and differential algebraic equations (DAE) [4]. The limit system as $\epsilon \to 0$ is the analog of an index 1 DAE, in which the algebraic equation is explicitly solved in terms of the differential variable. In the context of DAE, the initial condition that we called "well prepared" is called "consistent".

ii) This result does not guarantee the accuracy of the solution for the $m-n$ non-conserved quantities (sometimes refereed as the *algebraic variable*, by the analogy with differential algebraic systems). In fact, since the very last step in the scheme is not a projection toward the local equilibrium, a final layer effect occurs. The use of stiffly accurate schemes (i.e., schemes for which $a_{\nu j} = b_j, j = 1, \ldots, \nu$) in the implicit step may serve as a remedy to this problem. In order to obtain a uniformly accurate scheme even for the algebraic variable, more order conditions have to be imposed on the implicit scheme, which matches the numerical solution and the exact solution at various order in an expansion in the stiffness parameter in ϵ. A detailed analysis of this problem for IMEX Runge-Kutta schemes is reported in [9].

iii) The theorem guarantees that in the stiff limit the numerical scheme becomes the explicit RK scheme applied to the equilibrium system, and therefore the order of accuracy of the limiting scheme is greater than or equal to the order of accuracy of the original IMEX scheme (for the differential variables.)

6.4 Numerical Tests

We present some simple test cases that illustrate the behavior of the FD-WENO-RK-IMEX schemes. All computations have been performed by finite-difference WENO schemes with local Lax-Friedrichs flux and conservative variables. Of course the sharpness of the resolution of the numerical results can be improved using a less dissipative flux.

Most results are obtained with $N = 200$ grid points. The reference solution is computed on a much finer grid.

6.4.1 Broadwell Model

It is a simple model of the Boltzmann equation of gas dynamics [10, 12, 42].

The kinetic model is characterized by a hyperbolic system with relaxation of the form (6.1) with

$$u = (\rho,\, m,\, z), \quad F(U) = (m,\, z,\, m), \quad g(u) = \left(0,\, 0,\, \frac{1}{2}(\rho^2 + m^2 - 2\rho z)\right).$$

Here ε represents the mean free path of particles. The only conserved quantities are the density ρ and the momentum m.

In the fluid-dynamic limit $\varepsilon \to 0$ we have

$$z = z_E \equiv \frac{\rho^2 + m^2}{2\rho}, \tag{6.21}$$

and the Broadwell system is well approximated by the reduced system (6.19) with

$$u = (\rho,\, \rho v), \quad G(u) = \left(\rho v,\, \frac{1}{2}(\rho + \rho v^2)\right), \quad v = \frac{m}{\rho},$$

which represents the corresponding Euler equations of fluid dynamics.

Here we only test the shock-capturing properties of the schemes. Accuracy tests are reported in [49].

In particular, we consider non-smooth solutions characterized by the following two Riemann problems [12]:

$$\begin{aligned} \rho_l &= 2, & m_l &= 1, & z_l &= 1, & x &< 0.2, \\ \rho_r &= 1, & m_r &= 0.13962, & z_r &= 1, & x &> 0.2, \end{aligned} \tag{6.22}$$

$$\begin{aligned} \rho_l &= 1, & m_l &= 0, & z_l &= 1, & x &< 0, \\ \rho_r &= 0.2, & m_r &= 0, & z_r &= 1, & x &> 0. \end{aligned} \tag{6.23}$$

For brevity we report the numerical results obtained with the second-order IMEX-SSP2(2,2,2) and third-order IMEX-SSP3(4,3,3) schemes that we will refer to as IMEX-SSP2-WENO and IMEX-SSP3-WENO, respectively. The results are shown in Figures 6.1 and 6.2 for a Courant number $\Delta t/\Delta x = 0.5$. Both schemes, as expected, give an accurate description of the solution in all different regimes also using coarse meshes that do not resolve the small scales. In particular the shock formation in the fluid limit is well captured without spurious oscillations. We refer to [12, 29, 42, 46, 3] for a comparison of the present results with previous ones.

6.4.2 Shallow Water

First we consider a simple model of shallow water flow:

$$\begin{aligned} \partial_t h + \partial_x(hv) &= 0, \\ \partial_t(hv) + \partial_x(hv^2 + \frac{1}{2}h^2) &= \frac{h}{\epsilon}(\frac{h}{2} - v), \end{aligned} \tag{6.24}$$

where h is the water height with respect to the bottom, hv the flux, and the units are chosen so that the gravitational acceleration is $g = 1$. The source term is not realistic, and has been chosen for the purpose of checking the behavior of the schemes in the stiff regime. A similar model has been used as a test case by Shi-Jin [29].

The zero relaxation limit of the present model is given by the inviscid Burgers equation for h, while the velocity is algebraically related to h by $v = h/2$. In this limit this model is equivalent to the one used by Shi-Jin. The initial data we have considered is [29]

$$h = 1 + 0.2\sin(8\pi x), \quad hv = \frac{h^2}{2}, \tag{6.25}$$

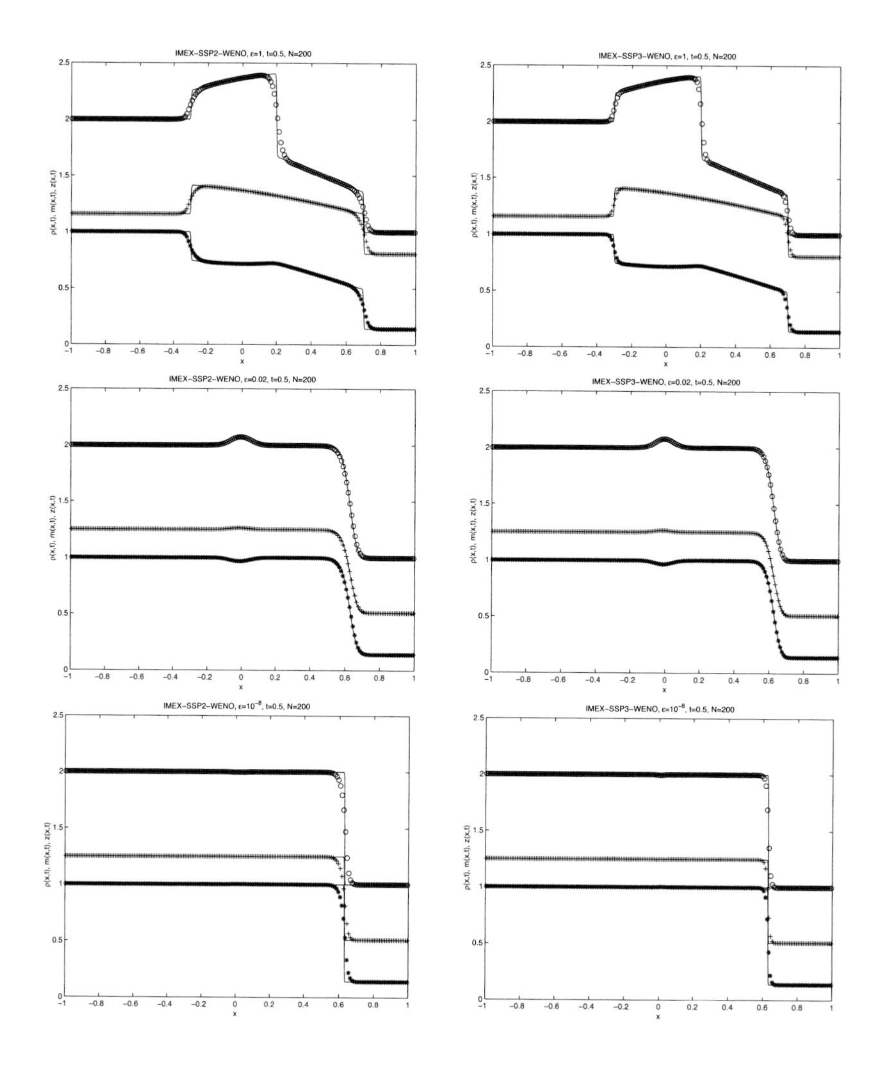

Figure 6.1: Numerical solution of the Broadwell equations with initial data (6.22) for $\rho(\circ)$, $m(*)$ and $z(+)$ at time $t = 0.5$. Left column IMEX-SSP2-WENO scheme, right column IMEX-SSP3-WENO scheme. From top to bottom, $\varepsilon = 1.0,\ 0.02,\ 10^{-8}$.

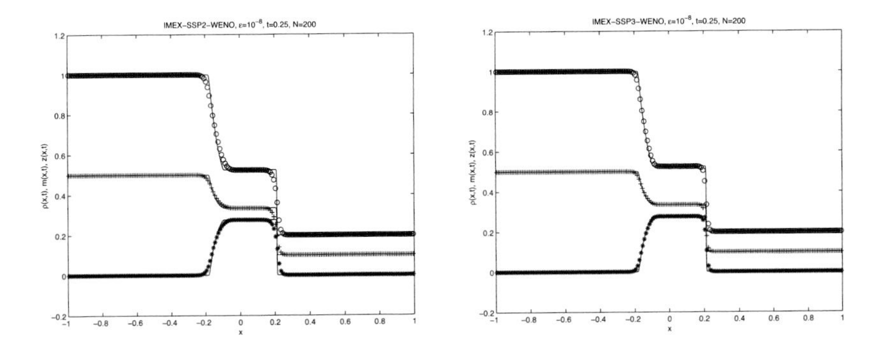

Figure 6.2: Numerical solution of the Broadwell equations with initial data (6.23) for $\rho(\circ)$, $m(*)$ and $z(+)$ at time $t = 0.25$ for $\varepsilon = 10^{-8}$. Left IMEX-SSP2-WENO scheme, right IMEX-SSP3-WENO scheme.

with $x \in [0, 1]$. The solution at $t = 0.5$ in the stiff regime $\epsilon = 10^{-8}$ using periodic boundary conditions is given in Figure 6.3. For IMEX-SSP2-WENO the dissipative effect due to the use of the Lax-Friedrichs flux is very pronounced. As expected, this effect becomes less relevant with the increase of the order of accuracy. We refer to [29] for a comparison with the present results.

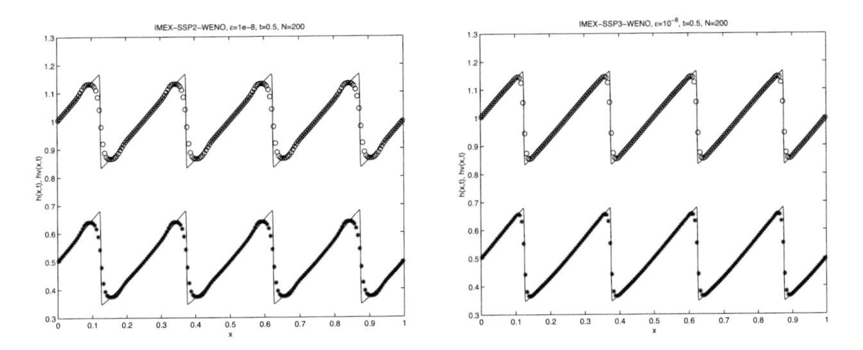

Figure 6.3: Numerical solution of the shallow water model with initial data (6.25) for $h(\circ)$ and $hv(*)$ at time $t = 0.5$ for $\varepsilon = 10^{-8}$. Left IMEX-SSP2-WENO scheme, right IMEX-SSP3-WENO scheme.

6.4.3 Traffic Flows

In [6] a new macroscopic model of vehicular traffic has been presented. The model consists of a continuity equation for the density ρ of vehicles together with an additional velocity equation that describes the mass flux variations due to the road conditions in front of the driver. The model can be written in conservative form as follows:

$$\begin{aligned} \partial_t \rho + \partial_x(\rho v) &= 0, \\ \partial_t(\rho w) + \partial_x(v\rho w) &= A\frac{\rho}{T}(V(\rho) - v), \end{aligned} \tag{6.26}$$

where $w = v + P(\rho)$ with $P(\rho)$ a given function describing the anticipation of road conditions in front of the drivers and $V(\rho)$ describing the dependence of the velocity with respect to the density for an equilibrium situation. The parameter T is the relaxation time and $A > 0$ is a positive constant.

If the relaxation time goes to zero, under the subcharacteristic condition

$$-P'(\rho) \leq V'(\rho) \leq 0, \quad \rho > 0,$$

we obtain the Lighthill-Whitham [64] model

$$\partial_t \rho + \partial_x(\rho V(\rho)) = 0. \tag{6.27}$$

A typical choice for the function $P(\rho)$ is given by

$$P(\rho) = \begin{cases} \dfrac{c_v}{\gamma}\left(\dfrac{\rho}{\rho_m}\right)^{\gamma} & \gamma > 0, \\ c_v \ln\left(\dfrac{\rho}{\rho_m}\right) & \gamma = 0, \end{cases}$$

where ρ_m is a given maximal density and c_v a constant with dimension of velocity. In our numerical results we assume $A = 1$ and an equilibrium velocity $V(\rho)$ fitting to experimental data [7]

$$V(\rho) = v_m \frac{\pi/2 + \arctan\left(\alpha\frac{\rho/\rho_m - \beta}{\rho/\rho_m - 1}\right)}{\pi/2 + \arctan(\alpha\beta)},$$

with $\alpha = 11$, $\beta = 0.22$ and v_m a maximal speed. We consider $\gamma = 0$ and, in order to fulfill the subcharacteristic condition, assume $c_v = 2$. All quantities are normalized so that $v_m = 1$ and $\rho_m = 1$. We consider a Riemann problem centered at $x = 0$ with left and right states

$$\rho_L = 0.05, \quad v_L = 0.05, \qquad \rho_R = 0.05, \quad v_R = 0.5. \tag{6.28}$$

The solution at $t = 1$ for $T = 0.2$ is given in Figure 6.4. The figure shows the development of the density of the vehicles. Both schemes give very similar results. Again, in the second-order scheme the shock is smeared out if compared to the third-order case. See [7] for more numerical results.

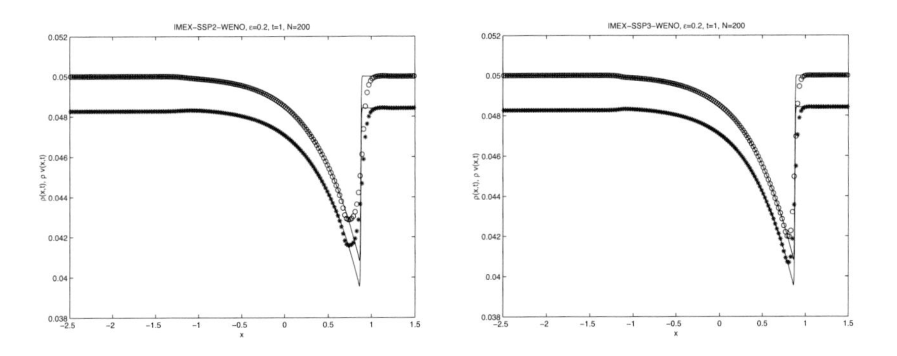

Figure 6.4: Numerical solution of the traffic model with initial data (6.28) for $\rho(\circ)$ and $\rho v(*)$ at time $t = 1$ for $\varepsilon = 0.2$. Left IMEX-SSP2-WENO scheme, right IMEX-SSP3-WENO scheme.

Appendix: Butcher Tableau of IMEX-RK

$$
\begin{array}{c|cc}
0 & 0 & 0 \\
1 & 1 & 0 \\
\hline
 & 1/2 & 1/2
\end{array}
\qquad
\begin{array}{c|cc}
\gamma & \gamma & 0 \\
1-\gamma & 1-2\gamma & \gamma \\
\hline
 & 1/2 & 1/2
\end{array}
\qquad
\gamma = 1 - \frac{1}{\sqrt{2}}
$$

Table A.1: Tableau for the explicit (left) implicit (right) IMEX-SSP2(2,2,2) L-stable scheme.

$$
\begin{array}{c|ccc}
0 & 0 & 0 & 0 \\
0 & 0 & 0 & 0 \\
1 & 0 & 1 & 0 \\
\hline
 & 0 & 1/2 & 1/2
\end{array}
\qquad
\begin{array}{c|ccc}
1/2 & 1/2 & 0 & 0 \\
0 & -1/2 & 1/2 & 0 \\
1 & 0 & 1/2 & 1/2 \\
\hline
 & 0 & 1/2 & 1/2
\end{array}
$$

Table A.2: Tableau for the explicit (left) implicit (right) IMEX-SSP2(3,2,2) stiffly accurate scheme.

$$
\begin{array}{c|ccc}
0 & 0 & 0 & 0 \\
1/2 & 1/2 & 0 & 0 \\
1 & 1/2 & 1/2 & 0 \\
\hline
 & 1/3 & 1/3 & 1/3
\end{array}
\qquad
\begin{array}{c|ccc}
1/4 & 1/4 & 0 & 0 \\
1/4 & 0 & 1/4 & 0 \\
1 & 1/3 & 1/3 & 1/3 \\
\hline
 & 1/3 & 1/3 & 1/3
\end{array}
$$

Table A.3: Tableau for the explicit (left) implicit (right) IMEX-SSP2(3,3,2) stiffly accurate scheme.

$$
\begin{array}{c|ccc}
0 & 0 & 0 & 0 \\
1 & 1 & 0 & 0 \\
1/2 & 1/4 & 1/4 & 0 \\
\hline
 & 1/6 & 1/6 & 2/3
\end{array}
\qquad
\begin{array}{c|ccc}
\gamma & \gamma & 0 & 0 \\
1-\gamma & 1-2\gamma & \gamma & 0 \\
1/2 & 1/2-\gamma & 0 & \gamma \\
\hline
 & 1/6 & 1/6 & 2/3
\end{array}
\qquad
\gamma = 1 - \frac{1}{\sqrt{2}}
$$

Table A.4: Tableau for the explicit (left) implicit (right) IMEX-SSP3(3,3,2) L-stable scheme.

$$
\begin{array}{c|cccc}
0 & 0 & 0 & 0 & 0 \\
0 & 0 & 0 & 0 & 0 \\
1 & 0 & 1 & 0 & 0 \\
1/2 & 0 & 1/4 & 1/4 & 0 \\
\hline
 & 0 & 1/6 & 1/6 & 2/3
\end{array}
\qquad
\begin{array}{c|cccc}
\alpha & \alpha & 0 & 0 & 0 \\
0 & -\alpha & \alpha & 0 & 0 \\
1 & 0 & 1-\alpha & \alpha & 0 \\
1/2 & \beta & \eta & 1/2-\beta-\eta-\alpha & \alpha \\
\hline
 & 0 & 1/6 & 1/6 & 2/3
\end{array}
$$

$$\alpha = 0.24169426078821, \quad \beta = 0.06042356519705 \quad \eta = 0.12915286960590$$

Table A.5: Tableau for the explicit (left) implicit (right) IMEX-SSP3(4,3,3) L-stable scheme.

Bibliography

[1] A.M. Anile and S. Pennisi, Thermodynamic derivation of the hydrodynamical model for charge transport in semiconductors. *Phys. Rev B* **46** (1992), 13186–13193.

[2] A.M. Anile and O. Muscato, Improved hydrodynamical model for carrier transport in semiconductors. *Phys. Rev B* **51** (1995), 16728–16740.

[3] M. Arora and P.L. Roe, Issues and strategies for hyperbolic problems with stiff source terms. In: *Barriers and challenges in computational fluid dynamics*, Hampton, VA, 1996, Kluwer Acad. Publ., Dordrecht, (1998), 139–154.

[4] U. Ascher and L. Petzold, *Computer Methods for Ordinary Differential Equations, and Differential Algebraic Equations.* SIAM, Philadelphia, 1998.

[5] U. Ascher, S. Ruuth and R.J. Spiteri, Implicit-explicit Runge-Kutta methods for time dependent Partial Differential Equations. *Appl. Numer. Math.* **25** (1997), 151–167.

[6] A.Aw and M. Rascle, Resurrection of second order models of traffic flow. *SIAM. J. Appl. Math.* **60** (2000), 916–938.

[7] A. Aw, A. Klar, T. Materne and M. Rascle, Derivation of continuum traffic flow models from microscopic follow the leader models. *SIAM J. Appl. Math.* **63** (2002), 259–278.

[8] F. Bianco, G. Puppo and G. Russo, High Order Central Schemes for Hyperbolic Systems of Conservation Laws. *SIAM J. Sci. Comp.* **21**, (1999), 294–322.

[9] S. Boscarino, Error analysis of IMEX Runge Kutta methods derived from Differential-Algebraic systems. *SIAM J. Numer. Anal.* **45**, (2007), 1600–1621.

[10] J.E. Broadwell, Shock structure in a simple discrete velocity gas. *Phys. Fluids* **7** (1964), 1013–1037.

[11] S. Chapman and T.G. Cowling, *The mathematical theory of nonuniform gases.* Cambridge University Press, London, 1960.

[12] R.E. Caflisch, S. Jin and G. Russo, Uniformly accurate schemes for hyperbolic systems with relaxation, *SIAM J. Numer. Anal.* **34** (1997), 246–281.

[13] G.Q. Chen, D. Levermore and T.P. Liu, Hyperbolic conservations laws with stiff relaxation terms and entropy, *Comm. Pure Appl. Math.* **47** (1994), 787–830.

[14] B. Cockburn, C. Johnson, C.-W. Shu and E. Tadmor, *Advanced Numerical Approximation of Nonlinear Hyperbolic Equations.* Lecture Notes in Mathematics (editor: A. Quarteroni), Springer, Berlin, 1998.

[15] R. Courant and K.O. Friedrichs, *Supersonic flow and shock waves.* Applied Mathematical Sciences, 21, New York, 1976.

[16] C. Dafermos, *Hyperbolic Conservation Laws in Continuum Physics*, Second Edition. Springer, Berlin–Heidelberg, 2005.

[17] B.O. Dia and M. Schatzman, Commutateur de certains semi-groupes holomorphes et applications aux directions alternées. *Mathematical Modelling and Numerical Analysis* **30** (1996), 343–383.

[18] B. Engquist and S. Osher, One sided difference approximations for nonlinear conservation laws. *Math. Comp.* **36** (1981), 321–351.

[19] K.O. Friedrichs and P.D. Lax, System of Conservation Laws With a Convex extension. *Proc. Nat. Acad. Sci. USA* **68** (1971), 1686–1688.

[20] R. Gatignol, *Théorie cinétique des gaz à répartition discrète de vitesses.* Lecture Notes in Physics, vol. 36, Springer, Berlin–New York, 1975.

[21] E. Godlewski, and P.-A. Raviart, *Numerical approximation of hyperbolic systems of conservation laws.* Applied Mathematical Sciences, 118, Springer, New York, 1996.

[22] S. Gottlieb and C.-W. Shu, Total Variation Diminishing Runge-Kutta schemes. *Math. Comp.* **67** (1998), 73–85.

[23] S. Gottlieb, C.-W. Shu and E. Tadmor, Strong-stability-preserving high order time discretization methods. *SIAM Reviews* **43** (2001), 89–112.

[24] E. Hairer and G. Wanner, *Solving ordinary differential equations*, Vol. 2: *Stiff and differential-algebraic problems.* Springer, New York, 1987.

[25] A. Harten, B. Engquist, S. Osher and S. Chakravarthy, Uniformly High Order Accurate Essentially Non-oscillatory Schemes III. *J. Comput. Phys.* **71** (1987), 231–303.

[26] A. Harten, P.D. Lax and B. van Leer, On upstream differencing and Godunov-type schemes for hyperbolic conservation laws. *SIAM Rev.* **25** (1983), 35–61.

[27] T. Jahnke and C. Lubich, *Error bounds for exponential operator splitting.* BIT, 2000, 735–744.

[28] G.-S. Jiang and C.-W. Shu, Efficient Implementation of Weighted ENO Schemes. *JCP* **126** (1996), 202–228.

[29] S. Jin, Runge-Kutta methods for hyperbolic systems with stiff relaxation terms. *J. Comp. Phys.* **122** (1995), 51–67.

[30] S. Jin and Z.P. Xin, The relaxation schemes for systems of conservation laws in arbitrary space dimensions. *Comm. Pure Appl. Math.* **48** no. 3 (1995), 235–276.

[31] C.A. Kennedy and M.H. Carpenter, Additive Runge-Kutta schemes for convection-diffusion-reaction equations. *Appl. Numer. Math.* **44** (2003), 139–181.

[32] R. Kupferman, A numerical study of the axisymmetric Couette-Taylor problem using a fast high-resolution second-order central scheme. *SIAM Journal on Scientific Computing* **20** (1998), 858–877.

[33] A. Kurganov and E. Tadmor, New High-Resolution Central Schemes for Nonlinear Conservation Laws and Convection-Diffusion Equations. *J. Comput. Phys.* **160** (2000), 214–282.

[34] A. Kurganov, S. Noelle and G.Petrova, Semidiscrete central-upwind schemes for hyperbolic conservation laws and Hamilton-Jacobi equations. *SIAM J. Sci. Comput.* **23** (2001), 707–740 .

[35] P.D. Lax, Weak Solutions of Non-Linear Hyperbolic Equations and Their Numerical Computation. *CPAM* **7** (1954), 159–193.

[36] P.D.Lax, *Hyperbolic systems of conservation laws and the mathematical theory of shock waves.* SIAM, Philadelphia, 1973.

[37] P.D. Lax and B. Wendroff, *Communications on Pure and Applied Mathematics*, 1960.

[38] R.J. LeVeque, *Numerical Methods for Conservation Laws.* Second edition. Lectures in Mathematics ETH Zürich, Birkhäuser, Basel, 1992.

[39] R.J. LeVeque, *Finite Volume methods for Hyperbolic Problems.* Cambridge University Press, 2002.

[40] D. Levy, G. Puppo and G. Russo, Central WENO Schemes for Hyperbolic Systems of Conservation Laws. *Math. Model. and Numer. Anal.* **33** no. 3 (1999), 547–571.

[41] D. Levy, G. Puppo and G. Russo, Compact Central WENO Schemes for Multidimensional Conservation Laws. *SIAM J. Sci. Comp.* **22** (2000), 656–672.

[42] S.F. Liotta, V. Romano and G. Russo, Central schemes for balance laws of relaxation type. *SIAM J. Numer. Anal.* **38** no. 4 (2000), 1337–1356.

[43] A. Marquina, Local piecewise hyperbolic reconstructions for nonlinear scalar conservation laws. *SIAM J. Sci. Comput.* **15** (1994), 892–915.

[44] I. Müller and T. Ruggeri, *Rational extended thermodynamics.* Springer, Berlin, 1998.

[45] H. Nessyahu and E. Tadmor, Non-oscillatory Central Differencing for Hyperbolic Conservation Laws. *J. Comput. Phys.* **87** no. 2 (1990), 408–463.

[46] L. Pareschi, Central differencing based numerical schemes for hyperbolic conservation laws with stiff relaxation terms. *SIAM J. Num. Anal.* **39** (2001), 1395–1417.

[47] L. Pareschi and G. Russo, Implicit-Explicit Runge-Kutta Schemes for Stiff Systems of Differential Equations. In: *Recent Trends in Numerical Analysis* (D. Trigiante, Ed.). Nova Science Publ., 2000, 269–288.

[48] L. Pareschi and G. Russo, High order asymptotically strong-stability-preserving methods for hyperbolic systems with stiff relaxation. In: *Proceedings HYP2002.* Pasadena USA, Springer, 2003, 241–255.

[49] L. Pareschi and G. Russo, Implicit-Explicit Runge-Kutta schemes and applications to hyperbolic systems with relaxation. *J. Sci. Comput.* **25** no. 1–2 (2005), 129–155.

[50] L. Pareschi, G. Puppo and G. Russo, Central Runge-Kutta schemes for conservation laws. *SIAM J. Sci. Comput.* **26** no. 3 (2005), 979–999.

[51] G. Puppo, Adaptive application of characteristic projection for central schemes. To appear in the proceedings of the HYP2002 conference.

[52] J. Qiu and C.-W. Shu, On the construction, comparison, and local characteristic decomposition for high-order central WENO schemes. *J. Comput. Phys.* **183** no. 1 (2002), 187–209.

[53] P.L. Roe, Approximate Riemann Solvers, Parameter Vectors, and Difference Schemes. *JCP* **43** (1981), 357–372.

[54] G. Russo, Central schemes and systems of balance laws. In: *Hyperbolic Partial Differential Equations, Theory, Numerics and Applications.* Vieweg, 2002.

[55] R. Sanders and W. Weiser, High resolution staggered mesh approach for nonlinear hyperbolic systems of conservation laws. *J. Comput. Phys.* **10** (1992), 314.

[56] C.-W. Shu, Essentially Non-Oscillatory and Weighted Essentially Non-Oscillatory Schemes for Hyperbolic Conservation Laws. In: *Advanced Numerical Approximation of Nonlinear Hyperbolic Equations.* Lecture Notes in Mathematics 1697 (editor: A. Quarteroni), Springer, Berlin, 1998.

[57] C.-W. Shu, Total variation diminishing time discretizations. *SIAM J. Sci. Stat. Comput.* **9** (1988), 1073–1084.

[58] C.-W. Shu and S. Osher, Efficient implementation of essentially nonoscillatory shock-capturing schemes. *J. Comput. Phys.* **77** no. 2 (1988), 439–471.

[59] C.-W. Shu and S. Osher, Efficient Implementation of Essentially Non-Oscillatory Shock-Capturing Schemes, II. *JCP* **83** (1989), 32–78.

[60] J. Shi, C. Hu and C.-W. Shu, A technique of treating negative weights in WENO schemes, *Journal of Computational Physics* **175** (2002) 108–127.

[61] R.J. Spiteri and S.J. Ruuth, A new class of optimal strong-stability-preserving time discretization methods. *SIAM. J. Num. Anal.* **40** no. 2 (2002), 469–491.

[62] G. Strang, On the construction and comparison of difference schemes. *SIAM J. Numer. Anal.* **5** (1968), 505–517.

[63] E. Tadmor, Approximate Solutions of Nonlinear Conservation Laws. In: Cockburn, Johnson, Shu and Tadmor Eds., Lecture Notes in Mathematics 1697, Springer, 1998.

[64] G.B. Whitham, *Linear and Nonlinear Waves.* Wiley, 1974.

[65] X. Zhong, Additive Semi-Implicit Runge-Kutta methods for computing high speed nonequilibrium reactive flows. *J. Comp. Phys.* **128** (1996), 19–31.

Part III

Discontinuous Galerkin Methods: General Approach and Stability

Chi-Wang Shu

Preface

In these lecture notes, we will give a general introduction to the discontinuous Galerkin (DG) methods for solving time-dependent, convection-dominated partial differential equations (PDEs), including the hyperbolic conservation laws, convection-diffusion equations, and PDEs containing higher-order spatial derivatives such as the KdV equations and other nonlinear dispersive wave equations. We will discuss cell entropy inequalities, nonlinear stability, and error estimates. The important ingredient of the design of DG schemes, namely the adequate choice of numerical fluxes, will be explained in detail. Issues related to the implementation of the DG method will also be addressed.

Chapter 1

Introduction

Discontinuous Galerkin (DG) methods are a class of finite-element methods using completely discontinuous basis functions, which are usually chosen as piecewise polynomials. Since the basis functions can be completely discontinuous, these methods have the flexibility which is not shared by typical finite-element methods, such as the allowance of arbitrary triangulation with hanging nodes, complete freedom in changing the polynomial degrees in each element independent of that in the neighbors (p adaptivity), and extremely local data structure (elements only communicate with immediate neighbors regardless of the order of accuracy of the scheme) and the resulting embarrassingly high parallel efficiency (usually more than 99% for a fixed mesh, and more than 80% for a dynamic load balancing with adaptive meshes which change often during time evolution), see, e.g. [5]. A very good example to illustrate the capability of the discontinuous Galerkin method in h-p adaptivity, efficiency in parallel dynamic load balancing, and excellent resolution properties is the successful simulation of the Rayleigh-Taylor flow instabilities in [38].

The first discontinuous Galerkin method was introduced in 1973 by Reed and Hill [37], in the framework of neutron transport, i.e., a time-independent linear hyperbolic equation. A major development of the DG method is carried out by Cockburn et al. in a series of papers [14, 13, 12, 10, 15], in which they have established a framework to easily solve *nonlinear* time-dependent problems, such as the Euler equations of gas dynamics, using explicit, nonlinearly stable high-order Runge-Kutta time discretizations [44] and DG discretization in space with exact or approximate Riemann solvers as interface fluxes and total variation bounded (TVB) nonlinear limiters [41] to achieve non-oscillatory properties for strong shocks.

The DG method has found rapid applications in such diverse areas as aeroacoustics, electro-magnetism, gas dynamics, granular flows, magneto-hydrodynamics, meteorology, modeling of shallow water, oceanography, oil recovery simulation, semiconductor device simulation, transport of contaminant in porous media, turbomachinery, turbulent flows, viscoelastic flows and weather forecasting, among many others. For more details, we refer to the survey paper [11], and other papers in that Springer volume, which contains the conference proceedings of the First International Symposium on Discontinuous Galerkin Methods held at Newport, Rhode Island in 1999. The lecture notes [8] is a good reference for many details, as well as the extensive review paper [17]. More recently, there are two special issues devoted to the discontinuous Galerkin method [18, 19], which contain many interesting papers in the development of the method in all aspects including algorithm design, analysis, implementation and applications.

Chapter 2

Time Discretization

In these lecture notes, we will concentrate on the method of lines DG methods, that is, we do not discretize the time variable. Therefore, we will briefly discuss the issue of time discretization at the beginning.

For hyperbolic problems or convection-dominated problems such as Navier-Stokes equations with high Reynolds numbers, we often use a class of high-order nonlinearly stable Runge-Kutta time discretizations. A distinctive feature of this class of time discretizations is that they are convex combinations of first-order forward Euler steps, hence they maintain strong stability properties in any semi-norm (total variation semi-norm, maximum norm, entropy condition, etc.) of the forward Euler step. Thus one only needs to prove nonlinear stability for the first-order forward Euler step, which is relatively easy in many situations (e.g., TVD schemes, see for example Section 3.2.2 below), and one automatically obtains the same strong stability property for the higher-order time discretizations in this class. These methods were first developed in [44] and [42], and later generalized in [20] and [21]. The most popular scheme in this class is the following third-order Runge-Kutta method for solving

$$u_t = L(u, t)$$

where $L(u, t)$ is a spatial discretization operator (it does not need to be, and often is not, linear!):

$$\begin{aligned} u^{(1)} &= u^n + \Delta t L(u^n, t^n), \\ u^{(2)} &= \frac{3}{4}u^n + \frac{1}{4}u^{(1)} + \frac{1}{4}\Delta t L(u^{(1)}, t^n + \Delta t), \\ u^{n+1} &= \frac{1}{3}u^n + \frac{2}{3}u^{(2)} + \frac{2}{3}\Delta t L(u^{(2)}, t^n + \frac{1}{2}\Delta t). \end{aligned} \tag{2.1}$$

Schemes in this class which are higher order or are of low storage also exist. For details, see the survey paper [43] and the review paper [21].

If the PDEs contain high-order spatial derivatives with coefficients not very small, then explicit time marching methods such as the Runge-Kutta methods described above suffer from severe time-step restrictions. It is an important and active research subject to study efficient time discretization for such situations, while still maintaining the advantages of the DG methods, such as their local nature and parallel efficiency. See, e.g. [46] for a study of several time discretization techniques for such situations. We will not further discuss this important issue though in these lectures.

Chapter 3

Discontinuous Galerkin Method for Conservation Laws

The discontinuous Galerkin method was first designed as an effective numerical method for solving hyperbolic conservation laws, which may have discontinuous solutions. In this section we will discuss the algorithm formulation, stability analysis, and error estimates for the discontinuous Galerkin method solving hyperbolic conservation laws.

3.1 Two-dimensional Steady-State Linear Equations

We now present the details of the original DG method in [37] for the two-dimensional steady-state linear convection equation

$$au_x + bu_y = f(x,y), \qquad 0 \le x,\ y \le 1, \tag{3.1}$$

where a and b are constants. Without loss of generality we assume $a > 0$, $b > 0$. The equation (3.1) is well posed when equipped with the inflow boundary condition

$$u(x,0) = g_1(x),\ \ 0 \le x \le 1 \qquad \text{and} \qquad u(0,y) = g_2(y),\ \ 0 \le y \le 1. \tag{3.2}$$

For simplicity, we assume a rectangular mesh to cover the computational domain $[0,1]^2$, consisting of cells

$$I_{i,j} = \left\{(x,y) : x_{i-\frac{1}{2}} \le x \le x_{i+\frac{1}{2}},\ y_{j-\frac{1}{2}} \le y \le y_{j+\frac{1}{2}}\right\}$$

for $1 \le i \le N_x$ and $1 \le j \le N_y$, where

$$0 = x_{\frac{1}{2}} < x_{\frac{3}{2}} < \cdots < x_{N_x+\frac{1}{2}} = 1$$

and

$$0 = y_{\frac{1}{2}} < y_{\frac{3}{2}} < \cdots < y_{N_y+\frac{1}{2}} = 1$$

are discretizations in x and y over $[0,1]$. We also denote

$$\Delta x_i = x_{i+\frac{1}{2}} - x_{i-\frac{1}{2}}, \ 1 \le i \le N_x; \qquad \Delta y_j = y_{j+\frac{1}{2}} - y_{j-\frac{1}{2}}, \ 1 \le j \le N_y;$$

and

$$h = \max\left(\max_{1\le i\le N_x} \Delta x_i, \ \max_{1\le j\le N_y} \Delta y_j \right).$$

We assume the mesh is regular, namely there is a constant $c > 0$ independent of h such that

$$\Delta x_i \ge ch, \ 1 \le i \le N_x; \qquad \Delta y_j \ge ch, \ 1 \le j \le N_y.$$

We define a finite-element space consisting of piecewise polynomials

$$V_h^k = \left\{ v : v|_{I_{i,j}} \in P^k(I_{i,j});\ 1 \le i \le N_x,\ 1 \le j \le N_y \right\}, \tag{3.3}$$

where $P^k(I_{i,j})$ denotes the set of polynomials of degree up to k defined on the cell $I_{i,j}$. Notice that functions in V_h^k may be discontinuous across cell interfaces.

The discontinuous Galerkin (DG) method for solving (3.1) is defined as follows: find the unique function $u_h \in V_h^k$ such that, for all test functions $v_h \in V_h^k$ and all $1 \le i \le N_x$ and $1 \le j \le N_y$, we have

$$\begin{aligned}
&- \int\int_{I_{i,j}} (a u_h (v_h)_x + b u_h (v_h)_y)\, dxdy + a \int_{y_{j-\frac{1}{2}}}^{y_{j+\frac{1}{2}}} \widehat{u}_h(x_{i+\frac{1}{2}}, y) v_h(x^-_{i+\frac{1}{2}}, y) dy \\
&- a \int_{y_{j-\frac{1}{2}}}^{y_{j+\frac{1}{2}}} \widehat{u}_h(x_{i-\frac{1}{2}}, y) v_h(x^+_{i-\frac{1}{2}}, y) dy + b \int_{x_{i-\frac{1}{2}}}^{x_{i+\frac{1}{2}}} \widehat{u}_h(x, y_{j+\frac{1}{2}}) v_h(x, y^-_{j+\frac{1}{2}}) dx \\
&- b \int_{x_{i-\frac{1}{2}}}^{x_{i+\frac{1}{2}}} \widehat{u}_h(x, y_{j-\frac{1}{2}}) v_h(x, y^+_{j-\frac{1}{2}}) dx = \int\int_{I_{i,j}} f\, v_h\, dxdy.
\end{aligned} \tag{3.4}$$

Here, $\widehat{u}_h$ is the so-called "numerical flux", which is a single-valued function defined at the cell interfaces and in general depending on the values of the numerical solution u_h from both sides of the interface, since u_h is discontinuous there. For the simple linear convection PDE (3.1), the numerical flux can be chosen according to the upwind principle, namely

$$\widehat{u}_h(x_{i+\frac{1}{2}}, y) = u_h(x^-_{i+\frac{1}{2}}, y), \qquad \widehat{u}_h(x, y_{j+\frac{1}{2}}) = u_h(x, y^-_{j+\frac{1}{2}}).$$

Notice that, for the boundary cell $i = 1$, the numerical flux for the left edge is defined using the given boundary condition

$$\widehat{u}_h(x_{\frac{1}{2}}, y) = g_2(y).$$

Likewise, for the boundary cell $j = 1$, the numerical flux for the bottom edge is defined by

$$\widehat{u}_h(x, y_{\frac{1}{2}}) = g_1(x).$$

We now look at the implementation of the scheme (3.4). If a local basis of $P^k(I_{i,j})$ is chosen and denoted as $\varphi^\ell_{i,j}(x,y)$ for $\ell = 1, 2, \ldots, K = (k+1)(k+2)/2$, we can express the numerical solution as

$$u_h(x,y) = \sum_{\ell=1}^{K} u^\ell_{i,j}\varphi^\ell_{i,j}(x,y), \qquad (x,y) \in I_{i,j},$$

and we should solve for the coefficients

$$u_{i,j} = \begin{pmatrix} u^1_{i,j} \\ \vdots \\ u^K_{i,j} \end{pmatrix},$$

which, according to the scheme (3.4), satisfies the linear equation

$$A_{i,j}u_{i,j} = rhs \tag{3.5}$$

where $A_{i,j}$ is a $K \times K$ matrix whose (ℓ, m)-th entry is given by

$$\begin{aligned} a^{\ell,m}_{i,j} = &-\int\int_{I_{i,j}} \left(a\varphi^m_{i,j}(x,y)(\varphi^\ell_{i,j}(x,y))_x + b\varphi^m_{i,j}(x,y)(\varphi^\ell_{i,j}(x,y))_y\right) dxdy \qquad (3.6)\\ &+ a\int_{y_{j-\frac{1}{2}}}^{y_{j+\frac{1}{2}}} \varphi^m_{i,j}(x_{i+\frac{1}{2}},y)\varphi^\ell_{i,j}(x_{i+\frac{1}{2}},y)dy \\ &+ b\int_{x_{i-\frac{1}{2}}}^{x_{i+\frac{1}{2}}} \varphi^m_{i,j}(x,y_{j+\frac{1}{2}})\varphi^\ell_{i,j}(x,y_{j+\frac{1}{2}})dx, \end{aligned}$$

and the ℓ-th entry of the right-hand side vector is given by

$$\begin{aligned} rhs^\ell = &\, a\int_{y_{j-\frac{1}{2}}}^{y_{j+\frac{1}{2}}} u_h(x^-_{i-\frac{1}{2}},y)\varphi^\ell_{i,j}(x_{i-\frac{1}{2}},y)dy + b\int_{x_{i-\frac{1}{2}}}^{x_{i+\frac{1}{2}}} u_h(x,y^-_{j-\frac{1}{2}})\varphi^\ell_{i,j}(x,y_{j-\frac{1}{2}})dx \\ &+ \int_{I_{i,j}} f\,\varphi^\ell_{i,j}\,dxdy, \end{aligned}$$

which depends on the information of u_h in the left cell $I_{i-1,j}$ and the bottom cell $I_{i,j-1}$, if they are in the computational domain, or on the boundary condition, if one or both of these cells are outside the computational domain. It is easy to verify that the matrix $A_{i,j}$ in (3.5) with entries given by (3.6) is invertible, hence the numerical solution u_h in the cell $I_{i,j}$ can be easily obtained by solving the small linear system (3.5), once the solution at the left and bottom cells $I_{i-1,j}$

and $I_{i,j-1}$ are already known, or if one or both of these cells are outside the computational domain. Therefore, we can obtain the numerical solution u_h in the following ordering: first we obtain it in the cell $I_{1,1}$, since both its left and bottom boundaries are equipped with the prescribed boundary conditions (3.2). We then obtain the solution in the cells $I_{2,1}$ and $I_{1,2}$. For $I_{2,1}$, the numerical solution u_h in its left cell $I_{1,1}$ is already available, and its bottom boundary is equipped with the prescribed boundary condition (3.2). Similar argument goes for the cell $I_{1,2}$. The next group of cells to be solved are $I_{3,1}$, $I_{2,2}$, $I_{1,3}$. It is clear that we can obtain the solution u_h sequentially in this way for all cells in the computational domain.

Clearly, this method does not involve any large system solvers and is very easy to implement. In [25], Lesaint and Raviart proved that this method is convergent with the optimal order of accuracy, namely $O(h^{k+1})$, in L^2-norm, when piecewise tensor product polynomials of degree k are used as basis functions. Numerical experiments indicate that the convergence rate is also optimal when the usual piecewise polynomials of degree k given by (3.3) are used instead.

Notice that, even though the method (3.4) is designed for the steady-state problem (3.1), it can be easily used on initial-boundary value problems of linear time-dependent hyperbolic equations: we just need to identify the time variable t as one of the spatial variables. It is also easily generalizable to higher dimensions.

The method described above can be easily designed and efficiently implemented on arbitrary triangulations. L^2-error estimates of $O(h^{k+1/2})$ where k is again the polynomial degree and h is the mesh size can be obtained when the solution is sufficiently smooth, for arbitrary meshes, see, e.g., [24]. This estimate is actually sharp for the most general situation [33], however in many cases the optimal $O(h^{k+1})$ error bound can be proved [39, 9]. In actual numerical computations, one almost always observes the optimal $O(h^{k+1})$ accuracy.

Unfortunately, even though the method (3.4) is easy to implement, accurate, and efficient, it cannot be easily generalized to linear systems, where the characteristic information comes from different directions, or to nonlinear problems, where the characteristic wind direction depends on the solution itself.

3.2 One-dimensional Time-dependent Conservation Laws

The difficulties mentioned at the end of the last subsection can be by-passed when the DG discretization is only used for the spatial variables, and the time discretization is achieved by explicit Runge-Kutta methods such as (2.1). This is the approach of the so-called Runge-Kutta discontinuous Galerkin (RKDG) method [14, 13, 12, 10, 15].

We start our discussion with the one-dimensional conservation law

$$u_t + f(u)_x = 0. \tag{3.7}$$

As before, we assume the following mesh to cover the computational domain $[0,1]$, consisting of cells $I_i = [x_{i-\frac{1}{2}}, x_{i+\frac{1}{2}}]$, for $1 \le i \le N$, where

$$0 = x_{\frac{1}{2}} < x_{\frac{3}{2}} < \cdots < x_{N+\frac{1}{2}} = 1.$$

We again denote

$$\Delta x_i = x_{i+\frac{1}{2}} - x_{i-\frac{1}{2}}, \quad 1 \le i \le N; \qquad h = \max_{1 \le i \le N} \Delta x_i.$$

We assume the mesh is regular, namely there is a constant $c > 0$ independent of h such that

$$\Delta x_i \ge ch, \qquad 1 \le i \le N.$$

We define a finite-element space consisting of piecewise polynomials

$$V_h^k = \left\{ v : v|_{I_i} \in P^k(I_i);\ 1 \le i \le N \right\}, \tag{3.8}$$

where $P^k(I_i)$ denotes the set of polynomials of degree up to k defined on the cell I_i. The semi-discrete DG method for solving (3.7) is defined as follows: find the unique function $u_h = u_h(t) \in V_h^k$ such that, for all test functions $v_h \in V_h^k$ and all $1 \le i \le N$, we have

$$\int_{I_i} (u_h)_t (v_h) dx - \int_{I_i} f(u_h)(v_h)_x dx + \widehat{f}_{i+\frac{1}{2}} v_h(x_{i+\frac{1}{2}}^-) - \widehat{f}_{i-\frac{1}{2}} v_h(x_{i-\frac{1}{2}}^+) = 0. \tag{3.9}$$

Here, $\widehat{f}_{i+\frac{1}{2}}$ is again the numerical flux, which is a single-valued function defined at the cell interfaces and in general depends on the values of the numerical solution u_h from both sides of the interface

$$\widehat{f}_{i+\frac{1}{2}} = \widehat{f}(u_h(x_{i+\frac{1}{2}}^-, t), u_h(x_{i+\frac{1}{2}}^+, t)).$$

We use the so-called monotone fluxes from finite-difference and finite-volume schemes for solving conservation laws, which satisfy the following conditions:

- Consistency: $\widehat{f}(u,u) = f(u)$.
- Continuity: $\widehat{f}(u^-, u^+)$ is at least Lipschitz continuous with respect to both arguments u^- and u^+.
- Monotonicity: $\widehat{f}(u^-, u^+)$ is a non-decreasing function of its first argument u^- and a non-increasing function of its second argument u^+. Symbolically $\widehat{f}(\uparrow, \downarrow)$.

Well-known monotone fluxes include the Lax-Friedrichs flux

$$\widehat{f}^{LF}(u^-, u^+) = \frac{1}{2}\left(f(u^-) + f(u^+) - \alpha(u^+ - u^-) \right), \qquad \alpha = \max_u |f'(u)|;$$

the Godunov flux

$$\widehat{f}^{God}(u^-, u^+) = \begin{cases} \min_{u^- \le u \le u^+} f(u), & \text{if } u^- < u^+, \\ \max_{u^+ \le u \le u^-} f(u), & \text{if } u^- \ge u^+; \end{cases}$$

and the Engquist-Osher flux

$$\widehat{f}^{EO} = \int_0^{u^-} \max(f'(u), 0) du + \int_0^{u^+} \min(f'(u), 0) du + f(0).$$

We refer to, e.g., [26] for more details about monotone fluxes.

3.2.1 Cell Entropy Inequality and L^2-Stability

It is well known that weak solutions of (3.7) may not be unique and the unique, physically relevant weak solution (the so-called entropy solution) satisfies the following entropy inequality

$$U(u)_t + F(u)_x \le 0 \tag{3.10}$$

in distribution sense, for any convex entropy $U(u)$ satisfying $U''(u) \ge 0$ and the corresponding entropy flux $F(u) = \int^u U'(u) f'(u) du$. It will be nice if a numerical approximation to (3.7) also shares a similar entropy inequality as (3.10). It is usually quite difficult to prove a discrete entropy inequality for finite-difference or finite-volume schemes, especially for high-order schemes and when the flux function $f(u)$ in (3.7) is not convex or concave, see, e.g., [28, 32]. However, it turns out that it is easy to prove that the DG scheme (3.9) satisfies a cell entropy inequality [23].

Proposition 3.1. *The solution u_h to the semi-discrete DG scheme (3.9) satisfies the following cell entropy inequality*

$$\frac{d}{dt} \int_{I_i} U(u_h)\, dx + \widehat{F}_{i+\frac{1}{2}} - \widehat{F}_{i-\frac{1}{2}} \le 0 \tag{3.11}$$

for the square entropy $U(u) = \frac{u^2}{2}$, for some consistent entropy flux

$$\widehat{F}_{i+\frac{1}{2}} = \widehat{F}(u_h(x^-_{i+\frac{1}{2}}, t), u_h(x^+_{i+\frac{1}{2}}, t))$$

satisfying $\widehat{F}(u, u) = F(u)$.

Proof. We introduce a short-hand notation

$$B_i(u_h; v_h) = \int_{I_i} (u_h)_t (v_h) dx - \int_{I_i} f(u_h)(v_h)_x dx \\ + \widehat{f}_{i+\frac{1}{2}} v_h(x^-_{i+\frac{1}{2}}) - \widehat{f}_{i-\frac{1}{2}} v_h(x^+_{i-\frac{1}{2}}). \tag{3.12}$$

If we take $v_h = u_h$ in the scheme (3.9), we obtain

$$B_i(u_h; u_h) = \int_{I_i} (u_h)_t (u_h) dx - \int_{I_i} f(u_h)(u_h)_x dx \\ + \widehat{f}_{i+\frac{1}{2}} u_h(x^-_{i+\frac{1}{2}}) - \widehat{f}_{i-\frac{1}{2}} u_h(x^+_{i-\frac{1}{2}}) = 0. \quad (3.13)$$

If we denote $\widetilde{F}(u) = \int^u f(u) du$, then (3.13) becomes

$$B_i(u_h; u_h) = \int_{I_i} U(u_h)_t dx - \widetilde{F}(u_h(x^-_{i+\frac{1}{2}})) \\ + \widetilde{F}(u_h(x^+_{i-\frac{1}{2}})) + \widehat{f}_{i+\frac{1}{2}} u_h(x^-_{i+\frac{1}{2}}) - \widehat{f}_{i-\frac{1}{2}} u_h(x^+_{i-\frac{1}{2}}) = 0,$$

or

$$B_i(u_h; u_h) = \int_{I_i} U(u_h)_t dx + \widehat{F}_{i+\frac{1}{2}} - \widehat{F}_{i-\frac{1}{2}} + \Theta_{i-\frac{1}{2}} = 0, \quad (3.14)$$

where

$$\widehat{F}_{i+\frac{1}{2}} = -\widetilde{F}(u_h(x^-_{i+\frac{1}{2}})) + \widehat{f}_{i+\frac{1}{2}} u_h(x^-_{i+\frac{1}{2}}), \quad (3.15)$$

and

$$\Theta_{i-\frac{1}{2}} = -\widetilde{F}(u_h(x^-_{i-\frac{1}{2}})) + \widehat{f}_{i-\frac{1}{2}} u_h(x^-_{i-\frac{1}{2}}) + \widetilde{F}(u_h(x^+_{i-\frac{1}{2}})) - \widehat{f}_{i-\frac{1}{2}} u_h(x^+_{i-\frac{1}{2}}). \quad (3.16)$$

It is easy to verify that the numerical entropy flux $\widehat{F}$ defined by (3.15) is consistent with the entropy flux $F(u) = \int^u U'(u) f'(u) du$ for $U(u) = \frac{u^2}{2}$. It is also easy to verify

$$\Theta = -\widetilde{F}(u_h^-) + \widehat{f} u_h^- + \widetilde{F}(u_h^+) - \widehat{f} u_h^+ = (u_h^+ - u_h^-)(\widetilde{F}'(\xi) - \widehat{f}) \geq 0,$$

where we have dropped the subscript $i - \frac{1}{2}$ since all quantities are evaluated there in $\Theta_{i-\frac{1}{2}}$. A mean value theorem is applied and ξ is a value between u^- and u^+, and we have used the fact $\widetilde{F}'(\xi) = f(\xi)$ and the monotonicity of the flux function $\widehat{f}$ to obtain the last inequality. This finishes the proof of the cell entropy inequality (3.11). □

We note that the proof does not depend on the accuracy of the scheme, namely it holds for the piecewise polynomial space (3.8) with any degree k. Also, the same proof can be given for the multi-dimensional DG scheme on any triangulation.

The cell entropy inequality trivially implies an L^2-stability of the numerical solution.

Proposition 3.2. *For periodic or compactly supported boundary conditions, the solution u_h to the semi-discrete* DG *scheme* (3.9) *satisfies the following L^2-stability*

$$\frac{d}{dt} \int_0^1 (u_h)^2 dx \leq 0, \quad (3.17)$$

or

$$\|u_h(\cdot,t)\| \le \|u_h(\cdot,0)\|. \tag{3.18}$$

Here and below, an unmarked norm is the usual L^2-norm.

Proof. We simply sum up the cell entropy inequality (3.11) over i. The flux terms telescope and there is no boundary term left because of the periodic or compact supported boundary condition. (3.17), and hence (3.18), are now immediate. □

Notice that both the cell entropy inequality (3.11) and the L^2-stability (3.17) are valid even when the exact solution of the conservation law (3.7) is discontinuous.

3.2.2 Limiters and Total Variation Stability

For discontinuous solutions, the cell entropy inequality (3.11) and the L^2-stability (3.17), although helpful, are not enough to control spurious numerical oscillations near discontinuities. In practice, especially for problems containing strong discontinuities, we often need to apply nonlinear limiters to control these oscillations and to obtain provable total variation stability.

For simplicity, we first consider the forward Euler time discretization of the semi-discrete DG scheme (3.9). Starting from a preliminary solution $u_h^{n,\mathrm{pre}} \in V_h^k$ at time level n (for the initial condition, $u_h^{0,\mathrm{pre}}$ is taken to be the L^2-projection of the analytical initial condition $u(\cdot,0)$ into V_h^k), we would like to "limit" or "preprocess" it to obtain a new function $u_h^n \in V_h^k$ before advancing it to the next time level: find $u_h^{n+1,\mathrm{pre}} \in V_h^k$ such that, for all test functions $v_h \in V_h^k$ and all $1 \le i \le N$, we have

$$\int_{I_i} \frac{u_h^{n+1,\mathrm{pre}} - u_h^n}{\Delta t} v_h dx - \int_{I_i} f(u_h^n)(v_h)_x dx + \widehat{f}^n_{i+\frac{1}{2}} v_h(x^-_{i+\frac{1}{2}}) - \widehat{f}^n_{i-\frac{1}{2}} v_h(x^+_{i-\frac{1}{2}}) = 0, \tag{3.19}$$

where $\Delta t = t^{n+1} - t^n$ is the time step. This limiting procedure to go from $u_h^{n,\mathrm{pre}}$ to u_h^n should satisfy the following two conditions:

- It should not change the cell averages of $u_h^{n,\mathrm{pre}}$. That is, the cell averages of u_h^n and $u_h^{n,\mathrm{pre}}$ are the same. This is for the conservation property of the DG method.

- It should not affect the accuracy of the scheme in smooth regions. That is, in the smooth regions this limiter does not change the solution, $u_h^n(x) = u_h^{n,\mathrm{pre}}(x)$.

There are many limiters discussed in the literature, and this is still an active research area, especially for multi-dimensional systems, see, e.g., [60]. We will only present an example [13] here.

We denote the cell average of the solution u_h as

$$\bar{u}_i = \frac{1}{\Delta x_i} \int_{I_i} u_h dx, \tag{3.20}$$

and we further denote

$$\widetilde{u}_i = u_h(x^-_{i+\frac{1}{2}}) - \bar{u}_i, \qquad \widetilde{\widetilde{u}}_i = \bar{u}_i - u_h(x^+_{i-\frac{1}{2}}). \tag{3.21}$$

The limiter should not change $\bar{u}_i$ but it may change $\widetilde{u}_i$ and/or $\widetilde{\widetilde{u}}_i$. In particular, the minmod limiter [13] changes $\widetilde{u}_i$ and $\widetilde{\widetilde{u}}_i$ into

$$\widetilde{u}_i^{(\mathrm{mod})} = m(\widetilde{u}_i, \Delta_+\bar{u}_i, \Delta_-\bar{u}_i), \qquad \widetilde{\widetilde{u}}_i^{(\mathrm{mod})} = m(\widetilde{\widetilde{u}}_i, \Delta_+\bar{u}_i, \Delta_-\bar{u}_i), \tag{3.22}$$

where

$$\Delta_+\bar{u}_i = \bar{u}_{i+1} - \bar{u}_i, \qquad \Delta_-\bar{u}_i = \bar{u}_i - \bar{u}_{i-1},$$

and the minmod function m is defined by

$$m(a_1, \cdots, a_\ell) = \begin{cases} s \min(|a_1|, \cdots, |a_\ell|), & \text{if } s = \mathrm{sign}(a_1) = \cdots \mathrm{sign}(a_\ell), \\ 0, & \text{otherwise.} \end{cases} \tag{3.23}$$

The limited function $u_h^{(\mathrm{mod})}$ is then recovered to maintain the old cell average (3.20) and the new point values given by (3.22), that is

$$u_h^{(\mathrm{mod})}(x^-_{i+\frac{1}{2}}) = \bar{u}_i + \widetilde{u}_i^{(\mathrm{mod})}, \qquad u_h^{(\mathrm{mod})}(x^+_{i-\frac{1}{2}}) = \bar{u}_i - \widetilde{\widetilde{u}}_i^{(\mathrm{mod})}, \tag{3.24}$$

by the definition (3.21). This recovery is unique for P^k polynomials with $k \leq 2$. For $k > 2$, we have extra freedom in obtaining $u_h^{(\mathrm{mod})}$. We could for example choose $u_h^{(\mathrm{mod})}$ to be the unique P^2 polynomial satisfying (3.20) and (3.24).

Before discussing the total variation stability of the DG scheme (3.19) with the pre-processing, we first present a simple lemma due to Harten [22].

Lemma 3.1 (Harten). *If a scheme can be written in the form*

$$u_i^{n+1} = u_i^n + C_{i+\frac{1}{2}}\Delta_+ u_i^n - D_{i-\frac{1}{2}}\Delta_- u_i^n \tag{3.25}$$

with periodic or compactly supported boundary conditions, where $C_{i+\frac{1}{2}}$ and $D_{i-\frac{1}{2}}$ may be nonlinear functions of the grid values u_j^n for $j = i-p, \ldots, i+q$ with some $p, q \geq 0$, satisfying

$$C_{i+\frac{1}{2}} \geq 0, \qquad D_{i+\frac{1}{2}} \geq 0, \qquad C_{i+\frac{1}{2}} + D_{i+\frac{1}{2}} \leq 1, \qquad \forall i, \tag{3.26}$$

then the scheme is TVD

$$TV(u^{n+1}) \leq TV(u^n),$$

where the total variation seminorm is defined by

$$TV(u) = \sum_i |\Delta_+ u_i|.$$

Proof. Taking the forward difference operation on (3.25) yields

$$\begin{aligned}\Delta_+ u_i^{n+1} &= \Delta_+ u_i^n + C_{i+\frac{3}{2}}\Delta_+ u_{i+1}^n - C_{i+\frac{1}{2}}\Delta_+ u_i^n - D_{i+\frac{1}{2}}\Delta_+ u_i^n + D_{i-\frac{1}{2}}\Delta_- u_i^n \\ &= (1 - C_{i+\frac{1}{2}} - D_{i+\frac{1}{2}})\Delta_+ u_i^n + C_{i+\frac{3}{2}}\Delta_+ u_{i+1}^n + D_{i-\frac{1}{2}}\Delta_- u_i^n.\end{aligned}$$

Thanks to (3.26) and using the periodic or compactly supported boundary condition, we can take the absolute value on both sides of the above equality and sum up over i to obtain

$$\begin{aligned}\sum_i |\Delta_+ u_i^{n+1}| \le& \sum_i (1 - C_{i+\frac{1}{2}} - D_{i+\frac{1}{2}})|\Delta_+ u_i^n| \\ &+ \sum_i C_{i+\frac{1}{2}}|\Delta_+ u_i^n| + \sum_i D_{i+\frac{1}{2}}|\Delta_+ u_i^n| = \sum_i |\Delta_+ u_i^n|.\end{aligned}$$

This finishes the proof. □

We define the "total variation in the means" semi-norm, or TVM, as

$$\mathrm{TVM}(u_h) = \sum_i |\Delta_+ \bar{u}_i|.$$

We then have the following stability result.

Proposition 3.3. *For periodic or compactly supported boundary conditions, the solution u_h^n of the DG scheme* (3.19), *with the "pre-processing" by the limiter, is total variation diminishing in the means* (TVDM), *that is*

$$\mathrm{TVM}(u_h^{n+1}) \le \mathrm{TVM}(u_h^n). \tag{3.27}$$

Proof. Taking $v_h = 1$ for $x \in I_i$ in (3.19) and dividing both sides by Δx_i, we obtain, by noticing (3.24),

$$\bar{u}_i^{n+1,\mathrm{pre}} = \bar{u}_i - \lambda_i \left(\widehat{f}(\bar{u}_i + \widetilde{u}_i, \bar{u}_{i+1} - \widetilde{\widetilde{u}}_{i+1}) - \widehat{f}(\bar{u}_{i-1} + \widetilde{u}_{i-1}, \bar{u}_i - \widetilde{\widetilde{u}}_i) \right),$$

where $\lambda_i = \frac{\Delta t}{\Delta x_i}$, and all quantities on the right-hand side are at the time level n. We can write the right hand side of the above equality in the Harten form (3.25) if we define $C_{i+\frac{1}{2}}$ and $D_{i-\frac{1}{2}}$ as follows

$$\begin{aligned} C_{i+\frac{1}{2}} &= -\lambda_i \frac{\widehat{f}(\bar{u}_i + \widetilde{u}_i, \bar{u}_{i+1} - \widetilde{\widetilde{u}}_{i+1}) - \widehat{f}(\bar{u}_i + \widetilde{u}_i, \bar{u}_i - \widetilde{\widetilde{u}}_i)}{\Delta_+ \bar{u}_i}, \\ D_{i-\frac{1}{2}} &= \lambda_i \frac{\widehat{f}(\bar{u}_i + \widetilde{u}_i, \bar{u}_i - \widetilde{\widetilde{u}}_i) - \widehat{f}(\bar{u}_{i-1} + \widetilde{u}_{i-1}, \bar{u}_i - \widetilde{\widetilde{u}}_i)}{\Delta_- \bar{u}_i}. \end{aligned} \tag{3.28}$$

We now need to verify that $C_{i+\frac{1}{2}}$ and $D_{i-\frac{1}{2}}$ defined in (3.28) satisfy (3.26). Indeed, we can write $C_{i+\frac{1}{2}}$ as

$$C_{i+\frac{1}{2}} = -\lambda_i \widehat{f}_2 \left[1 - \frac{\widetilde{\widetilde{u}}_{i+1}}{\Delta_+ \bar{u}_i} + \frac{\widetilde{\widetilde{u}}_i}{\Delta_+ \bar{u}_i} \right], \tag{3.29}$$

in which $\widehat{f}_2$ is defined as

$$\widehat{f}_2 = \frac{\widehat{f}(\bar{u}_i + \widetilde{u}_i, \bar{u}_{i+1} - \widetilde{\widetilde{u}}_{i+1}) - \widehat{f}(\bar{u}_i + \widetilde{u}_i, \bar{u}_i - \widetilde{\widetilde{u}}_i)}{(\bar{u}_{i+1} - \widetilde{\widetilde{u}}_{i+1}) - (\bar{u}_i - \widetilde{\widetilde{u}}_i)},$$

and hence

$$0 \le -\lambda_i \widehat{f}_2 = -\lambda_i \frac{\widehat{f}(\bar{u}_i + \widetilde{u}_i, \bar{u}_{i+1} - \widetilde{\widetilde{u}}_{i+1}) - \widehat{f}(\bar{u}_i + \widetilde{u}_i, \bar{u}_i - \widetilde{\widetilde{u}}_i)}{(\bar{u}_{i+1} - \widetilde{\widetilde{u}}_{i+1}) - (\bar{u}_i - \widetilde{\widetilde{u}}_i)} \le \lambda_i L_2, \quad (3.30)$$

where we have used the monotonicity and Lipschitz continuity of $\widehat{f}$, and L_2 is the Lipschitz constant of $\widehat{f}$ with respect to its second argument. Also, since u_h^n is the pre-processed solution by the minmod limiter, $\widetilde{\widetilde{u}}_{i+1}$ and $\widetilde{\widetilde{u}}_i$ are the modified values defined by (3.22), hence

$$0 \le \frac{\widetilde{\widetilde{u}}_{i+1}}{\Delta_+ \bar{u}_i} \le 1, \qquad 0 \le \frac{\widetilde{\widetilde{u}}_i}{\Delta_+ \bar{u}_i} \le 1. \quad (3.31)$$

Therefore, we have, by (3.29), (3.30) and (3.31),

$$0 \le C_{i+\frac{1}{2}} \le 2\lambda_i L_2.$$

Similarly, we can show that

$$0 \le D_{i+\frac{1}{2}} \le 2\lambda_{i+1} L_1$$

where L_1 is the Lipschitz constant of $\widehat{f}$ with respect to its first argument. This proves (3.26) if we take the time step so that

$$\lambda \le \frac{1}{2(L_1 + L_2)}$$

where $\lambda = \max_i \lambda_i$. The TVDM property (3.27) then follows from the Harten Lemma and the fact that the limiter does not change cell averages, hence $\text{TVM}(u_h^{n+1}) = \text{TVM}(u_h^{n+1,pre})$. □

Even though the previous proposition is proved only for the first-order Euler forward time discretization, the special TVD (or strong stability preserving, SSP) Runge-Kutta time discretizations [44, 21] allow us to obtain the same stability result for the fully discretized RKDG schemes.

Proposition 3.4. *Under the same conditions as those in Proposition* 3.3, *the solution* u_h^n *of the* DG *scheme* (3.19), *with the Euler forward time discretization replaced by any* SSP *Runge-Kutta time discretization* [21] *such as* (2.1), *is* TVDM. □

We still need to verify that the limiter (3.22) does not affect accuracy in smooth regions. If u_h is an approximation to a (locally) smooth function u, then a simple Taylor expansion gives

$$\widetilde{u}_i = \frac{1}{2} u_x(x_i)\Delta x_i + O(h^2), \qquad \widetilde{\widetilde{u}}_i = \frac{1}{2} u_x(x_i)\Delta x_i + O(h^2),$$

while

$$\Delta_+ \bar{u}_i = \frac{1}{2} u_x(x_i)(\Delta x_i + \Delta x_{i+1}) + O(h^2), \quad \Delta_- \bar{u}_i = \frac{1}{2} u_x(x_i)(\Delta x_i + \Delta x_{i-1}) + O(h^2).$$

Clearly, when we are in a smooth and monotone region, namely when $u_x(x_i)$ is away from zero, the first argument in the minmod function (3.22) is of the same sign as the second and third arguments and is smaller in magnitude (for a uniform mesh it is about half of their magnitude), when h is small. Therefore, since the minmod function (3.23) picks the smallest argument (in magnitude) when all the arguments are of the same sign, the modified values $\widetilde{u}_i^{(\mathrm{mod})}$ and $\widetilde{\widetilde{u}}_i^{(\mathrm{mod})}$ in (3.22) will take the unmodified values $\widetilde{u}_i$ and $\widetilde{\widetilde{u}}_i$, respectively. That is, the limiter does not affect accuracy in smooth, monotone regions.

On the other hand, the TVD limiter (3.22) does kill accuracy at smooth extrema. This is demonstrated by numerical results and is a consequence of the general results about TVD schemes, that they are at most second-order accurate for smooth but non-monotone solutions [31]. Therefore, in practice we often use a total variation bounded (TVB) corrected limiter

$$\widetilde{m}(a_1, \cdots, a_\ell) = \begin{cases} a_1, & \text{if } |a_1| \le Mh^2, \\ m(a_1, \ldots, a_\ell), & \text{otherwise}, \end{cases}$$

instead of the original minmod function (3.23), where the TVB parameter M has to be chosen adequately [13]. The DG scheme would then be total variation bounded in the means (TVBM) and uniformly high-order accurate for smooth solutions. We will not discuss more details here and refer the readers to [13].

We would like to remark that the limiters discussed in this subsection were first used for finite-volume schemes [30]. When discussing limiters, the DG methods and finite-volume schemes have many similarities.

3.2.3 Error Estimates for Smooth Solutions

If we assume the exact solution of (3.7) is smooth, we can obtain optimal L^2-error estimates. Such error estimates can be obtained for the general nonlinear conservation law (3.7) and for fully discretized RKDG methods, see [58]. However, for simplicity we will give here the proof only for the semi-discrete DG scheme and the linear version of (3.7):

$$u_t + u_x = 0, \tag{3.32}$$

for which the monotone flux is taken as the simple upwind flux $\widehat{f}(u^-, u^+) = u^-$. Of course the proof is the same for $u_t + au_x = 0$ with any constant a.

Proposition 3.5. *The solution u_h of the* DG *scheme* (3.9) *for the* PDE (3.32) *with a smooth solution u satisfies the error estimate*

$$\|u - u_h\| \leq Ch^{k+1} \tag{3.33}$$

where C depends on u and its derivatives but is independent of h.

Proof. The DG scheme (3.9), when using the notation in (3.12), can be written as

$$B_i(u_h; v_h) = 0, \tag{3.34}$$

for all $v_h \in V_h$ and for all i. It is easy to verify that the exact solution of the PDE (3.32) also satisfies

$$B_i(u; v_h) = 0, \tag{3.35}$$

for all $v_h \in V_h$ and for all i. Subtracting (3.34) from (3.35) and using the linearity of B_i with respect to its first argument, we obtain the error equation

$$B_i(u - u_h; v_h) = 0, \tag{3.36}$$

for all $v_h \in V_h$ and for all i.

We now define a special projection P into V_h. For a given smooth function w, the projection Pw is the unique function in V_h which satisfies, for each i,

$$\int_{I_i} (Pw(x) - w(x))v_h(x)dx = 0 \qquad \forall v_h \in P^{k-1}(I_i); \qquad Pw(x^-_{i+\frac{1}{2}}) = w(x_{i+\frac{1}{2}}). \tag{3.37}$$

Standard approximation theory [7] implies, for a smooth function w,

$$\|Pw(x) - w(x)\| \leq Ch^{k+1} \tag{3.38}$$

where here and below C is a generic constant depending on w and its derivatives but independent of h (which may not have the same value in different places). In particular, in (3.38), $C = \widetilde{C}\|w\|_{H^{k+1}}$ where $\|w\|_{H^{k+1}}$ is the standard Sobolev $(k+1)$ norm and $\widetilde{C}$ is a constant independent of w.

We now take:

$$v_h = Pu - u_h \tag{3.39}$$

in the error equation (3.36), and denote

$$e_h = Pu - u_h, \qquad \varepsilon_h = u - Pu \tag{3.40}$$

to obtain

$$B_i(e_h; e_h) = -B_i(\varepsilon_h; e_h). \tag{3.41}$$

For the left-hand side of (3.41), we use the cell entropy inequality (see (3.14)) to obtain

$$B_i(e_h; e_h) = \frac{1}{2}\frac{d}{dt}\int_{I_i} (e_h)^2 dx + \widehat{F}_{i+\frac{1}{2}} - \widehat{F}_{i-\frac{1}{2}} + \Theta_{i-\frac{1}{2}}, \tag{3.42}$$

where $\Theta_{i-\frac{1}{2}} \geq 0$. As to the right-hand side of (3.41), we first write out all the terms

$$-B_i(\varepsilon_h; e_h) = -\int_{I_i} (\varepsilon_h)_t e_h dx + \int_{I_i} \varepsilon_h (e_h)_x dx - (\varepsilon_h)^-_{i+\frac{1}{2}} (e_h)^-_{i+\frac{1}{2}} + (\varepsilon_h)^-_{i-\frac{1}{2}} (e_h)^+_{i+\frac{1}{2}}.$$

Noticing the properties (3.37) of the projection P, we have

$$\int_{I_i} \varepsilon_h (e_h)_x dx = 0$$

because $(e_h)_x$ is a polynomial of degree at most $k-1$, and

$$(\varepsilon_h)^-_{i+\frac{1}{2}} = u_{i+\frac{1}{2}} - (Pu)^-_{i+\frac{1}{2}} = 0$$

for all i. Therefore, the right-hand side of (3.41) becomes

$$-B_i(\varepsilon_h; e_h) = -\int_{I_i} (\varepsilon_h)_t e_h dx \leq \frac{1}{2} \left(\int_{I_i} ((\varepsilon_h)_t)^2 dx + \int_{I_i} (e_h)^2 dx \right). \tag{3.43}$$

Plugging (3.42) and (3.43) into the equality (3.41), summing up over i, and using the approximation result (3.38), we obtain

$$\frac{d}{dt} \int_0^1 (e_h)^2 dx \leq \int_0^1 (e_h)^2 dx + Ch^{2k+2}.$$

A Gronwall's inequality, the fact that the initial error

$$\|u(\cdot, 0) - u_h(\cdot, 0)\| \leq Ch^{k+1}$$

(usually the initial condition $u_h(\cdot, 0)$ is taken as the L^2-projection of the analytical initial condition $u(\cdot, 0)$), and the approximation result (3.38) finally give us the error estimate (3.33). □

3.3 Comments for Multi-dimensional Cases

Even though we have only discussed the two-dimensional steady-state and one-dimensional time-dependent cases in previous subsections, most of the results also hold for multi-dimensional cases with arbitrary triangulations. For example, the semi-discrete DG method for the two-dimensional time-dependent conservation law

$$u_t + f(u)_x + g(u)_y = 0 \tag{3.44}$$

is defined as follows. The computational domain is partitioned into a collection of cells $\triangle_i$, which in 2D could be rectangles, triangles, etc., and the numerical solution is a polynomial of degree k in each cell $\triangle_i$. The degree k could change with the cell, and there is no continuity requirement of the two polynomials along an interface of two cells. Thus, instead of only one degree of freedom per cell as in a finite-volume scheme, namely the cell average of the solution, there are now $K = \frac{(k+1)(k+2)}{2}$ degrees of freedom per cell for a DG method using piecewise k-th degree polynomials in 2D. These K degrees of freedom are chosen as the coefficients of the polynomial when expanded in a local basis. One could use a locally orthogonal basis to simplify the computation, but this is not essential.

The DG method is obtained by multiplying (3.44) by a test function $v(x,y)$ (which is also a polynomial of degree k in the cell), integrating over the cell $\triangle_j$, and integrating by parts:

$$\frac{d}{dt}\int_{\triangle_j} u(x,y,t)v(x,y)dxdy - \int_{\triangle_j} F(u)\cdot\nabla v\, dxdy + \int_{\partial\triangle_j} F(u)\cdot n\, v\, ds = 0, \quad (3.45)$$

where $F = (f,g)$, and n is the outward unit normal of the cell boundary $\partial\triangle_j$. The line integral in (3.45) is typically discretized by a Gaussian quadrature of sufficiently high order of accuracy,

$$\int_{\partial\triangle_j} F\cdot n\, v\, ds \approx |\partial\triangle_j| \sum_{k=1}^{q} \omega_k F(u(G_k,t))\cdot n\, v(G_k),$$

where $F(u(G_k,t))\cdot n$ is replaced by a numerical flux (approximate or exact Riemann solvers). For scalar equations the numerical flux can be taken as any of the monotone fluxes discussed in Section 3.2 along the normal direction of the cell boundary. For example, one could use the simple Lax-Friedrichs flux, which is given by

$$F(u(G_k,t))\cdot n \approx \frac{1}{2}\left[\left(F(u^-(G_k,t))+F(u^+(G_k,t))\right)\cdot n - \alpha\left(u^+(G_k,t)-u^-(G_k,t)\right)\right],$$

where α is taken as an upper bound for the eigenvalues of the Jacobian in the n direction, and u^- and u^+ are the values of u inside the cell $\triangle_j$ and outside the cell $\triangle_j$ (inside the neighboring cell) at the Gaussian point G_k. $v(G_k)$ is taken as $v^-(G_k)$, namely the value of v inside the cell $\triangle_j$ at the Gaussian point G_k. The volume integral term $\int_{\triangle_j} F(u)\cdot\nabla v\, dxdy$ can be computed either by a numerical quadrature or by a quadrature free implementation [2] for special systems such as the compressible Euler equations. Notice that if a locally orthogonal basis is chosen, the time derivative term $\frac{d}{dt}\int_{\triangle_j} u(x,y,t)v(x,y)dxdy$ would be explicit and there is no mass matrix to invert. However, even if the local basis is not orthogonal, one still only needs to invert a small $K\times K$ local mass matrix (by hand) and there is never a global mass matrix to invert as in a typical finite-element method.

For scalar equations (3.44), the cell entropy inequality described in Proposition 3.1 holds for arbitrary triangulation. The limiter described in Section 3.2.2 can also be defined for arbitrary triangulation, see [10]. Instead of the TVDM property given in Proposition 3.3, for multi-dimensional cases one can prove the maximum norm stability of the limited scheme, see [10]. The optimal error estimate given in Proposition 3.5 can be proved for tensor product meshes and basis functions, and for certain specific triangulations when the usual piecewise k-th degree polynomial approximation spaces are used [39, 9]. For the most general cases, an L^2-error estimate of half an order lower $O(h^{k+\frac{1}{2}})$ can be proved [24], which is actually sharp [33].

For nonlinear hyperbolic equations including symmetrizable systems, if the solution of the PDE is smooth, L^2-error estimates of $O(h^{k+1/2} + \Delta t^2)$ where Δt is the time step can be obtained for the fully discrete Runge-Kutta discontinuous Galerkin method with second-order Runge-Kutta time discretization. For upwind fluxes the optimal $O(h^{k+1} + \Delta t^2)$ error estimate can be obtained. See [58, 59].

As an example of the excellent numerical performance of the RKDG scheme, we show in Figures 3.1 and 3.2 the solution of the second order (piecewise linear) and seventh order (piecewise polynomial of degree 6) DG methods for the linear transport equation

$$u_t + u_x = 0, \qquad \text{or} \qquad u_t + u_x + u_y = 0,$$

on the domain $(0, 2\pi) \times (0, T)$ or $(0, 2\pi)^2 \times (0, T)$ with the characteristic function of the interval $(\frac{\pi}{2}, \frac{3\pi}{2})$ or the square $(\frac{\pi}{2}, \frac{3\pi}{2})^2$ as initial condition and periodic boundary conditions [17]. Notice that the solution is for a *very long* time, $t = 100\pi$ (50 time periods), with a relatively coarse mesh. We can see that the second-order scheme smears the fronts, however the seventh-order scheme maintains the shape of the solution almost as well as the initial condition! The excellent performance can be achieved by the DG method on multi-dimensional linear systems using unstructured meshes, hence it is a very good method for solving, e.g. Maxwell equations of electromagnetism and linearized Euler equations of aeroacoustics.

To demonstrate that the DG method also works well for nonlinear systems, we show in Figure 3.3 the DG solution of the forward facing step problem by solving the compressible Euler equations of gas dynamics [15]. We can see that the roll-ups of the contact line caused by a physical instability are resolved well, especially by the third-order DG scheme.

In summary, we can say the following about the discontinuous Galerkin methods for conservation laws:

1. They can be used for arbitrary triangulation, including those with hanging nodes. Moreover, the degree of the polynomial, hence the order of accuracy, in each cell can be independently decided. Thus the method is ideally suited for *h*-*p* (mesh size and order of accuracy) refinements and adaptivity.

2. The methods have excellent parallel efficiency. Even with space time adaptivity and load balancing the parallel efficiency can still be over 80%, see [38].

3. They should be the methods of choice if geometry is complicated or if adaptivity is important, especially for problems with long time evolution of smooth solutions.

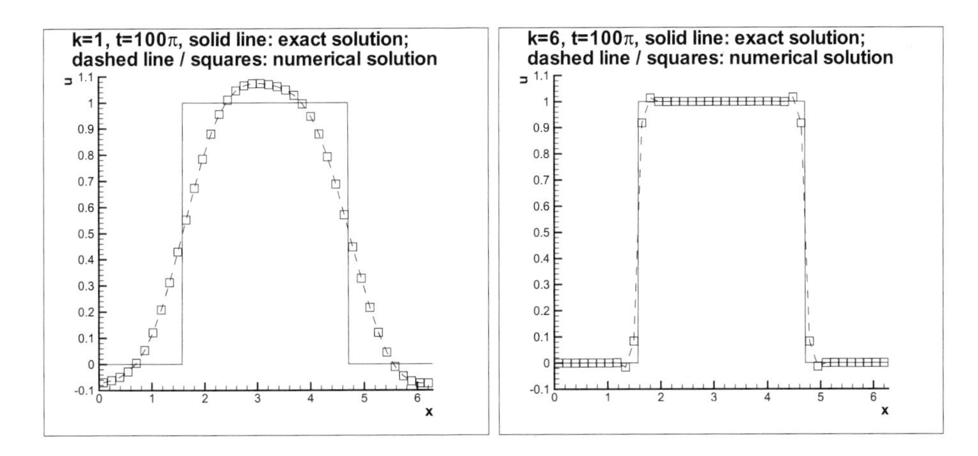

Figure 3.1: Transport equation: Comparison of the exact and the RKDG solutions at $T = 100\pi$ with second order (P^1, left) and seventh order (P^6, right) RKDG methods. One-dimensional results with 40 cells, exact solution (solid line) and numerical solution (dashed line and symbols, one point per cell).

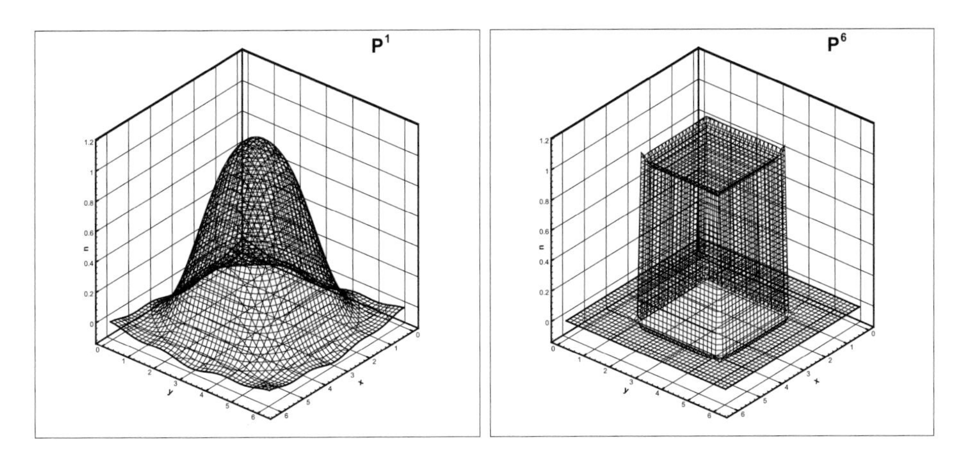

Figure 3.2: Transport equation: Comparison of the exact and the RKDG solutions at $T = 100\pi$ with second order (P^1, left) and seventh order (P^6, right) RKDG methods. Two-dimensional results with 40×40 cells.

4. For problems containing strong shocks, the nonlinear limiters are still less robust than the advanced WENO philosophy. There is a parameter (the TVB constant) for the user to tune for each problem, see [13, 10, 15]. For rectangular meshes the limiters work better than for triangular ones. In recent years, WENO based limiters have been investigated [35, 34, 36].

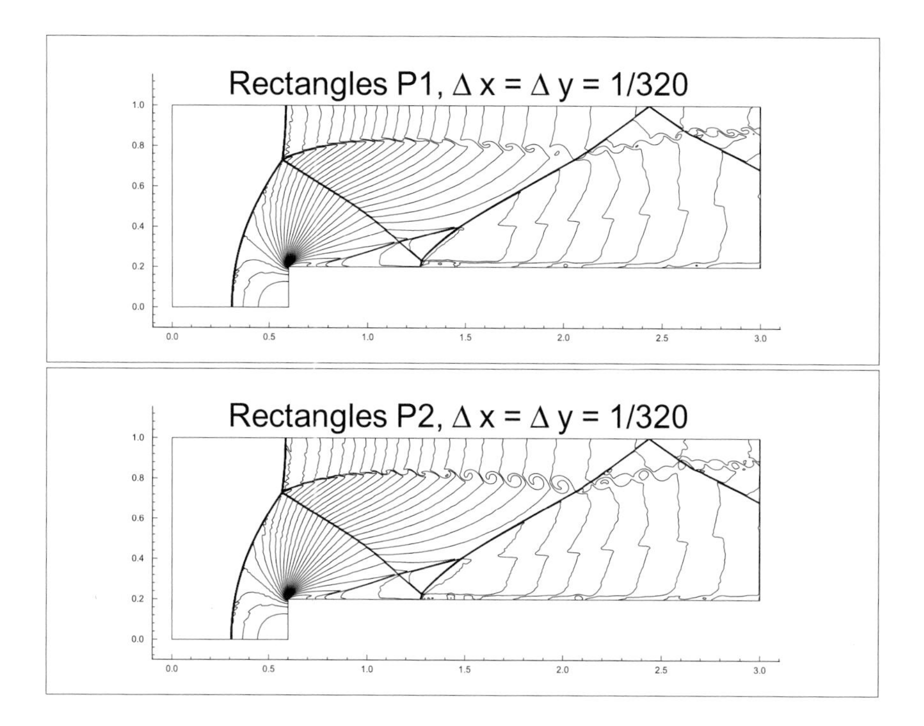

Figure 3.3: Forward facing step. Zoomed-in region. $\Delta x = \Delta y = \frac{1}{320}$. Top: P^1 elements; bottom: P^2 elements.

Chapter 4

Discontinuous Galerkin Method for Convection-Diffusion Equations

In this section we discuss the discontinuous Galerkin method for time-dependent convection-diffusion equations

$$u_t + \sum_{i=1}^{d} f_i(u)_{x_i} - \sum_{i=1}^{d}\sum_{j=1}^{d}(a_{ij}(u)u_{x_j})_{x_i} = 0, \tag{4.1}$$

where $(a_{ij}(u))$ is a symmetric, semi-positive definite matrix. There are several different formulations of discontinuous Galerkin methods for solving such equations, e.g., [1, 4, 6, 29, 45], however in this section we will only discuss the local discontinuous Galerkin (LDG) method [16].

For equations containing higher-order spatial derivatives, such as the convection-diffusion equation (4.1), discontinuous Galerkin methods cannot be directly applied. This is because the solution space, which consists of piecewise polynomials discontinuous at the element interfaces, is not regular enough to handle higher derivatives. This is a typical "non-conforming" case in finite elements. A naive and careless application of the discontinuous Galerkin method directly to the heat equation containing second derivatives could yield a method which behaves nicely in the computation but is "inconsistent" with the original equation and has $O(1)$ errors to the exact solution [17, 57].

The idea of local discontinuous Galerkin methods for time-dependent partial differential equations with higher derivatives, such as the convection-diffusion equation (4.1), is to rewrite the equation into a first-order system, then apply the discontinuous Galerkin method on the system. A key ingredient for the success of such methods is the correct design of interface numerical fluxes. These fluxes must be designed to guarantee stability and local solvability of all the auxiliary variables introduced to approximate the derivatives of the solution. The local solvability of all the auxiliary variables is why the method is called a "local" discontinuous Galerkin method in [16].

The first local discontinuous Galerkin method was developed by Cockburn and Shu [16], for the convection-diffusion equation (4.1) containing second derivatives. Their work was motivated by the successful numerical experiments of Bassi and Rebay [3] for the compressible Navier-Stokes equations.

In the following we will discuss the stability and error estimates for the LDG method for convection-diffusion equations. We present details only for the one-dimensional case and will mention briefly the generalization to multi-dimensions in Section 4.4.

4.1 LDG Scheme Formulation

We consider the one-dimensional convection-diffusion equation

$$u_t + f(u)_x = (a(u)u_x)_x \tag{4.2}$$

with $a(u) \geq 0$. We rewrite this equation as the system

$$u_t + f(u)_x = (b(u)q)_x, \qquad q - B(u)_x = 0, \tag{4.3}$$

where

$$b(u) = \sqrt{a(u)}, \qquad B(u) = \int^u b(u)du. \tag{4.4}$$

The finite-element space is still given by (3.8). The semi-discrete LDG scheme is defined as follows. Find $u_h, q_h \in V_h^k$ such that, for all test functions $v_h, p_h \in V_h^k$ and all $1 \leq i \leq N$, we have

$$\begin{aligned} &\int_{I_i} (u_h)_t (v_h) dx - \int_{I_i} (f(u_h) - b(u_h) q_h)(v_h)_x dx \\ &\qquad + (\widehat{f} - \widehat{b}\widehat{q})_{i+\frac{1}{2}} (v_h)^-_{i+\frac{1}{2}} - (\widehat{f} - \widehat{b}\widehat{q})_{i-\frac{1}{2}} (v_h)^+_{i-\frac{1}{2}} = 0, \\ &\int_{I_i} q_h p_h dx + \int_{I_i} B(u_h)(p_h)_x dx - \widehat{B}_{i+\frac{1}{2}} (p_h)^-_{i+\frac{1}{2}} + \widehat{B}_{i-\frac{1}{2}} (p_h)^+_{i-\frac{1}{2}} = 0. \end{aligned} \tag{4.5}$$

Here, all the "hat" terms are the numerical fluxes, namely single-valued functions defined at the cell interfaces which typically depend on the discontinuous numerical solution from both sides of the interface. We already know from Section 3 that the convection flux $\widehat{f}$ should be chosen as a monotone flux. However, the upwinding principle is no longer a valid guiding principle for the design of the diffusion fluxes $\widehat{b}$, $\widehat{q}$ and $\widehat{B}$. In [16], sufficient conditions for the choices of these diffusion fluxes to guarantee the stability of the scheme (4.5) are given. Here, we will discuss a particularly attractive choice, called "alternating fluxes", defined as

$$\widehat{b} = \frac{B(u_h^+) - B(u_h^-)}{u_h^+ - u_h^-}, \qquad \widehat{q} = q_h^+, \qquad \widehat{B} = B(u_h^-). \tag{4.6}$$

The important point is that $\widehat{q}$ and $\widehat{B}$ should be chosen from different directions. Thus, the choice

$$\widehat{b} = \frac{B(u_h^+) - B(u_h^-)}{u_h^+ - u_h^-}, \qquad \widehat{q} = q_h^-, \qquad \widehat{B} = B(u_h^+)$$

is also fine.

Notice that, from the second equation in the scheme (4.5), we can solve q_h explicitly and locally (in cell I_i) in terms of u_h, by inverting the small mass matrix inside the cell I_i. This is why the method is referred to as the "local" discontinuous Galerkin method.

4.2 Stability Analysis

Similar to the case for hyperbolic conservation laws, we have the following "cell entropy inequality" for the LDG method (4.5).

Proposition 4.1. *The solution u_h, q_h to the semi-discrete* LDG *scheme* (4.5) *satisfies the following "cell entropy inequality"*

$$\frac{1}{2}\frac{d}{dt}\int_{I_i} (u_h)^2\, dx + \int_{I_i} (q_h)^2 dx + \widehat{F}_{i+\frac{1}{2}} - \widehat{F}_{i-\frac{1}{2}} \leq 0 \tag{4.7}$$

for some consistent entropy flux

$$\widehat{F}_{i+\frac{1}{2}} = \widehat{F}(u_h(x^-_{i+\frac{1}{2}}, t), q_h(x^-_{i+\frac{1}{2}}, t); u_h(x^+_{i+\frac{1}{2}}, t), q_h(x^+_{i+\frac{1}{2}}))$$

satisfying $\widehat{F}(u,u) = F(u) - ub(u)q$ where, as before, $F(u) = \int^u uf'(u)du$.

Proof. We introduce a short-hand notation

$$\begin{aligned} B_i(u_h, q_h; v_h, p_h) = {} & \int_{I_i} (u_h)_t (v_h) dx - \int_{I_i} (f(u_h) - b(u_h)q_h)(v_h)_x dx \\ & + (\widehat{f} - \widehat{bq})_{i+\frac{1}{2}} (v_h)^-_{i+\frac{1}{2}} - (\widehat{f} - \widehat{bq})_{i-\frac{1}{2}} (v_h)^+_{i-\frac{1}{2}} \\ & + \int_{I_i} q_h p_h dx + \int_{I_i} B(u_h)(p_h)_x dx - \widehat{B}_{i+\frac{1}{2}} (p_h)^-_{i+\frac{1}{2}} + \widehat{B}_{i-\frac{1}{2}} (p_h)^+_{i-\frac{1}{2}}. \end{aligned} \tag{4.8}$$

If we take $v_h = u_h$, $p_h = q_h$ in the scheme (4.5), we obtain

$$B_i(u_h, q_h; u_h, q_h) = \int_{I_i} (u_h)_t (u_h) dx \tag{4.9}$$

$$- \int_{I_i} (f(u_h) - b(u_h)q_h)(u_h)_x dx$$

$$+ (\widehat{f} - \widehat{b}\widehat{q})_{i+\frac{1}{2}} (u_h)^-_{i+\frac{1}{2}} - (\widehat{f} - \widehat{b}\widehat{q})_{i-\frac{1}{2}} (u_h)^+_{i-\frac{1}{2}} \tag{4.10}$$

$$+ \int_{I_i} (q_h)^2 dx + \int_{I_i} B(u_h)(q_h)_x dx - \widehat{B}_{i+\frac{1}{2}} (q_h)^-_{i+\frac{1}{2}} + \widehat{B}_{i-\frac{1}{2}} (q_h)^+_{i-\frac{1}{2}}$$

$$= 0.$$

If we denote $\widetilde{F}(u) = \int^u f(u) du$, then (4.9) becomes

$$B_i(u_h, q_h; u_h, q_h) = \frac{1}{2} \frac{d}{dt} \int_{I_i} (u_h)^2 \, dx + \int_{I_i} (q_h)^2 dx$$
$$+ \widehat{F}_{i+\frac{1}{2}} - \widehat{F}_{i-\frac{1}{2}} + \Theta_{i-\frac{1}{2}} = 0, \tag{4.11}$$

where

$$\widehat{F} = -\widetilde{F}(u_h^-) + \widehat{f} u_h^- - \widehat{b} q_h^+ u_h^- \tag{4.12}$$

and

$$\Theta = -\widetilde{F}(u_h^-) + \widehat{f} u_h^- + \widetilde{F}(u_h^+) - \widehat{f} u_h^+, \tag{4.13}$$

where we have used the definition of the numerical fluxes (4.6). Notice that we have omitted the subindex $i - \frac{1}{2}$ in the definitions of $\widehat{F}$ and Θ. It is easy to verify that the numerical entropy flux $\widehat{F}$ defined by (4.12) is consistent with the entropy flux $F(u) - ub(u)q$. As Θ in (4.13) is the same as that in (3.16) for the conservation law case, we readily have $\Theta \geq 0$. This finishes the proof of (4.7). □

We again note that the proof does not depend on the accuracy of the scheme, namely it holds for the piecewise polynomial space (3.8) with any degree k. Also, the same proof can be given for multi-dimensional LDG schemes on any triangulation.

As before, the cell entropy inequality trivially implies an L^2-stability of the numerical solution.

Proposition 4.2. *For periodic or compactly supported boundary conditions, the solution* u_h, q_h *to the semi-discrete* LDG *scheme* (4.5) *satisfies the following* L^2-*stability*

$$\frac{d}{dt} \int_0^1 (u_h)^2 dx + 2 \int_0^1 (q_h)^2 dx \leq 0, \tag{4.14}$$

or

$$\|u_h(\cdot, t)\| + 2 \int_0^t \|q_h(\cdot, \tau)\| d\tau \leq \|u_h(\cdot, 0)\|. \tag{4.15}$$

□

Notice that both the cell entropy inequality (4.7) and the L^2-stability (4.14) are valid regardless of whether the convection-diffusion equation (4.2) is convection-dominated or diffusion-dominated and regardless of whether the exact solution is smooth or not. The diffusion coefficient $a(u)$ can be degenerate (equal to zero) in any part of the domain. The LDG method is particularly attractive for convection-dominated convection-diffusion equations, when traditional continuous finite-element methods are less stable.

4.3 Error Estimates

Again, if we assume the exact solution of (4.2) is smooth, we can obtain optimal L^2-error estimates. Such error estimates can be obtained for the general nonlinear convection-diffusion equation (4.2), see [53]. However, for simplicity we will give here the proof only for the heat equation:

$$u_t = u_{xx} \tag{4.16}$$

defined on $[0, 1]$ with periodic boundary conditions.

Proposition 4.3. *The solution u_h and q_h to the semi-discrete* DG *scheme* (4.5) *for the* PDE (4.16) *with a smooth solution u satisfies the error estimate*

$$\int_0^1 (u(x,t) - u_h(x,t))^2\, dx + \int_0^t \int_0^1 (u_x(x,\tau) - q_h(x,\tau))^2\, dx d\tau \leq C h^{2(k+1)}, \tag{4.17}$$

where C depends on u and its derivatives but is independent of h.

Proof. The DG scheme (4.5), when using the notation in (4.8), can be written as

$$B_i(u_h, q_h; v_h, p_h) = 0, \tag{4.18}$$

for all $v_h, p_h \in V_h$ and for all i. It is easy to verify that the exact solution u and $q = u_x$ of the PDE (4.16) also satisfies

$$B_i(u, q; v_h, p_h) = 0, \tag{4.19}$$

for all $v_h, p_h \in V_h$ and for all i. Subtracting (4.18) from (4.19) and using the linearity of B_i with respect to its first two arguments, we obtain the error equation

$$B_i(u - u_h, q - q_h; v_h, p_h) = 0, \tag{4.20}$$

for all $v_h, p_h \in V_h$ and for all i.

Recall the special projection P defined in (3.37). We also define another special projection Q as follows. For a given smooth function w, the projection Qw is the unique function in V_h which satisfies, for each i,

$$\int_{I_i} (Qw(x) - w(x)) v_h(x) dx = 0 \qquad \forall v_h \in P^{k-1}(I_i); \qquad Qw(x^+_{i-\frac{1}{2}}) = w(x_{i-\frac{1}{2}}). \tag{4.21}$$

Similar to P, we also have, by standard approximation theory [7], that

$$\|Qw(x) - w(x)\| \le Ch^{k+1}, \qquad \forall w \in H^{k+1}(\Omega), \tag{4.22}$$

where C is a constant depending on w and its derivatives but independent of h.

We now take

$$v_h = Pu - u_h, \qquad p_h = Qq - q_h \tag{4.23}$$

in the error equation (4.20), and denote

$$e_h = Pu - u_h, \;\; \bar{e}_h = Qq - q_h; \qquad \varepsilon_h = u - Pu, \;\; \bar{\varepsilon}_h = q - Qq, \tag{4.24}$$

to obtain

$$B_i(e_h, \bar{e}_h; e_h, \bar{e}_h) = -B_i(\varepsilon_h, \bar{\varepsilon}_h; e_h, \bar{e}_h). \tag{4.25}$$

For the left-hand side of (4.25), we use the cell entropy inequality (see (4.11)) to obtain

$$B_i(e_h, \bar{e}_h; e_h, \bar{e}_h) = \frac{1}{2}\frac{d}{dt}\int_{I_i}(e_h)^2 dx + \int_{I_i}(\bar{e}_h)^2 dx + \widehat{F}_{i+\frac{1}{2}} - \widehat{F}_{i-\frac{1}{2}} + \Theta_{i-\frac{1}{2}}, \tag{4.26}$$

where $\Theta_{i-\frac{1}{2}} \ge 0$ (in fact we can easily verify, from (4.13), that $\Theta_{i-\frac{1}{2}} = 0$ for the special case of the heat equation (4.16)). As to the right-hand side of (4.25), we first write out all the terms

$$\begin{aligned}
-B_i(\varepsilon_h, \bar{\varepsilon}_h; e_h, \bar{e}_h) = &-\int_{I_i}(\varepsilon_h)_t e_h dx \\
&-\int_{I_i}\bar{\varepsilon}_h (e_h)_x dx + (\bar{\varepsilon}_h)^+_{i+\frac{1}{2}}(e_h)^-_{i+\frac{1}{2}} - (\bar{\varepsilon}_h)^+_{i-\frac{1}{2}}(e_h)^+_{i-\frac{1}{2}} \\
&-\int_{I_i}\bar{\varepsilon}_h \bar{e}_h dx \\
&-\int_{I_i}\varepsilon_h (\bar{e}_h)_x dx + (\varepsilon_h)^-_{i+\frac{1}{2}}(\bar{e}_h)^-_{i+\frac{1}{2}} - (\varepsilon_h)^-_{i-\frac{1}{2}}(\bar{e}_h)^+_{i-\frac{1}{2}}.
\end{aligned}$$

Noticing the properties (3.37) and (4.21) of the projections P and Q, we have

$$\int_{I_i}\bar{\varepsilon}_h (e_h)_x dx = 0, \qquad \int_{I_i}\varepsilon_h (\bar{e}_h)_x dx = 0,$$

because $(e_h)_x$ and $(\bar{e}_h)_x$ are polynomials of degree at most $k-1$, and

$$(\varepsilon_h)^-_{i+\frac{1}{2}} = u_{i+\frac{1}{2}} - (Pu)^-_{i+\frac{1}{2}} = 0, \qquad (\bar{\varepsilon}_h)^+_{i+\frac{1}{2}} = q_{i+\frac{1}{2}} - (Qq)^+_{i+\frac{1}{2}} = 0,$$

for all i. Therefore, the right-hand side of (4.25) becomes

$$\begin{aligned}
-B_i(\varepsilon_h, \bar{\varepsilon}_h; e_h, \bar{e}_h) &= -\int_{I_i}(\varepsilon_h)_t e_h dx - \int_{I_i}\bar{\varepsilon}_h \bar{e}_h dx \\
&\le \frac{1}{2}\left(\int_{I_i}((\varepsilon_h)_t)^2 dx + \int_{I_i}(e_h)^2 dx + \int_{I_i}(\bar{\varepsilon}_h)^2 dx + \int_{I_i}(\bar{e}_h)^2 dx\right).
\end{aligned} \tag{4.27}$$

Plugging (4.26) and (4.27) into the equality (4.25), summing up over i, and using the approximation results (3.38) and (4.22), we obtain

$$\frac{d}{dt}\int_0^1 (e_h)^2 dx + \int_0^1 (\bar{e}_h)^2 dx \leq \int_0^1 (e_h)^2 dx + Ch^{2k+2}.$$

A Gronwall's inequality, the fact that the initial error

$$\|u(\cdot, 0) - u_h(\cdot, 0)\| \leq Ch^{k+1}$$

and the approximation results (3.38) and (4.22) finally give us the error estimate (4.17). □

4.4 Multi-Dimensions

Even though we have only discussed one-dimensional cases in this section, the algorithm and its analysis can be easily generalized to the multi-dimensional equation (4.1). The stability analysis is the same as for the one-dimensional case in Section 4.2. The optimal $O(h^{k+1})$ error estimates can be obtained on tensor product meshes and polynomial spaces, along the same line as that in Section 4.3. For general triangulations and piecewise polynomials of degree k, a sub-optimal error estimate of $O(h^k)$ can be obtained. We will not provide the details here and refer to [16, 53].

Chapter 5

Discontinuous Galerkin Method for PDEs Containing Higher-Order Spatial Derivatives

We now consider the DG method for solving PDEs containing higher-order spatial derivatives. Even though there are other possible DG schemes for such PDEs, e.g. those designed in [6], we will only discuss the local discontinuous Galerkin (LDG) method in this section.

5.1 LDG Scheme for the KdV Equations

We first consider PDEs containing third spatial derivatives. These are usually nonlinear dispersive wave equations, for example the following general KdV-type equations

$$u_t + \sum_{i=1}^{d} f_i(u)_{x_i} + \sum_{i=1}^{d} \Big(r_i'(u) \sum_{j=1}^{d} g_{ij}(r_i(u)_{x_i})_{x_j} \Big)_{x_i} = 0, \tag{5.1}$$

where $f_i(u)$, $r_i(u)$ and $g_{ij}(q)$ are arbitrary (smooth) nonlinear functions. The one-dimensional KdV equation

$$u_t + (\alpha u + \beta u^2)_x + \sigma u_{xxx} = 0, \tag{5.2}$$

where α, β and σ are constants, is a special case of the general class (5.1).

Stable LDG schemes for solving (5.1) were first designed in [55]. We will concentrate our discussion on the one-dimensional case. For the one-dimensional generalized KdV-type equations

$$u_t + f(u)_x + (r'(u)g(r(u)_x)_x)_x = 0, \tag{5.3}$$

where $f(u)$, $r(u)$ and $g(q)$ are arbitrary (smooth) nonlinear functions, the LDG method is based on rewriting it as the following system,

$$u_t + (f(u) + r'(u)p)_x = 0, \qquad p - g(q)_x = 0, \qquad q - r(u)_x = 0. \tag{5.4}$$

The finite-element space is still given by (3.8). The semi-discrete LDG scheme is defined as follows. Find $u_h, p_h, q_h \in V_h^k$ such that, for all test functions $v_h, w_h, z_h \in V_h^k$ and all $1 \le i \le N$, we have

$$\begin{aligned} &\int_{I_i} (u_h)_t (v_h) dx - \int_{I_i} (f(u_h) + r'(u_h)p_h)(v_h)_x dx \\ &\quad + (\widehat{f} + \widehat{r'}\widehat{p})_{i+\frac{1}{2}} (v_h)^-_{i+\frac{1}{2}} - (\widehat{f} + \widehat{r'}\widehat{p})_{i-\frac{1}{2}} (v_h)^+_{i-\frac{1}{2}} = 0, \\ &\int_{I_i} p_h w_h dx + \int_{I_i} g(q_h)(w_h)_x dx - \widehat{g}_{i+\frac{1}{2}} (w_h)^-_{i+\frac{1}{2}} + \widehat{g}_{i-\frac{1}{2}} (w_h)^+_{i-\frac{1}{2}} = 0, \\ &\int_{I_i} q_h z_h dx + \int_{I_i} r(u_h)(z_h)_x dx - \widehat{r}_{i+\frac{1}{2}} (z_h)^-_{i+\frac{1}{2}} + \widehat{r}_{i-\frac{1}{2}} (z_h)^+_{i-\frac{1}{2}} = 0. \end{aligned} \tag{5.5}$$

Here again, all the "hat" terms are the numerical fluxes, namely single-valued functions defined at the cell interfaces which typically depend on the discontinuous numerical solution from both sides of the interface. We already know from Section 3 that the convection flux $\widehat{f}$ should be chosen as a monotone flux. It is important to design the other fluxes suitably in order to guarantee stability of the resulting LDG scheme. In fact, the upwinding principle is still a valid guiding principle here, since the KdV-type equation (5.3) is a dispersive wave equation for which waves are propagating with a direction. For example, the simple linear equation

$$u_t + u_{xxx} = 0,$$

which corresponds to (5.3) with $f(u) = 0$, $r(u) = u$ and $g(q) = q$, admits the following simple wave solution

$$u(x,t) = \sin(x+t),$$

that is, information propagates from right to left. This motivates the following choice of numerical fluxes, discovered in [55]:

$$\widehat{r'} = \frac{r(u_h^+) - r(u_h^-)}{u_h^+ - u_h^-}, \qquad \widehat{p} = p_h^+, \qquad \widehat{g} = \widehat{g}(q_h^-, q_h^+), \qquad \widehat{r} = r(u_h^-). \tag{5.6}$$

Here, $-\widehat{g}(q_h^-, q_h^+)$ is a monotone flux for $-g(q)$, namely $\widehat{g}$ is a non-increasing function in the first argument and a non-decreasing function in the second argument. The important point is again the "alternating fluxes", namely $\widehat{p}$ and $\widehat{r}$ should come from opposite sides. Thus

$$\widehat{r'} = \frac{r(u_h^+) - r(u_h^-)}{u_h^+ - u_h^-}, \qquad \widehat{p} = p_h^-, \qquad \widehat{g} = \widehat{g}(q_h^-, q_h^+), \qquad \widehat{r} = r(u_h^+)$$

would also work.

Notice that, from the third equation in the scheme (5.5), we can solve q_h explicitly and locally (in cell I_i) in terms of u_h, by inverting the small mass matrix inside the cell I_i. Then, from the second equation in the scheme (5.5), we can solve p_h explicitly and locally (in cell I_i) in terms of q_h. Thus only u_h is the global unknown and the auxiliary variables q_h and p_h can be solved in terms of u_h locally. This is why the method is referred to as the "local" discontinuous Galerkin method.

5.1.1 Stability Analysis

Similar to the case for hyperbolic conservation laws and convection-diffusion equations, we have the following "cell entropy inequality" for the LDG method (5.5).

Proposition 5.1. *The solution u_h to the semi-discrete* LDG *scheme* (5.5) *satisfies the following "cell entropy inequality"*

$$\frac{1}{2}\frac{d}{dt}\int_{I_i} (u_h)^2\, dx + \widehat{F}_{i+\frac{1}{2}} - \widehat{F}_{i-\frac{1}{2}} \leq 0 \tag{5.7}$$

for some consistent entropy flux

$$\widehat{F}_{i+\frac{1}{2}} = \widehat{F}(u_h(x_{i+\frac{1}{2}}^-, t), p_h(x_{i+\frac{1}{2}}^-, t), q_h(x_{i+\frac{1}{2}}^-, t); u_h(x_{i+\frac{1}{2}}^+, t), p_h(x_{i+\frac{1}{2}}^+, t), q_h(x_{i+\frac{1}{2}}^+))$$

satisfying $\widehat{F}(u, u) = F(u) + ur'(u)p - G(q)$ where $F(u) = \int^u uf'(u)du$ and $G(q) = \int^q qg(q)dq$.

Proof. We introduce a short-hand notation

$$\begin{aligned} B_i(u_h,p_h,q_h;v_h,w_h,z_h) = &\int_{I_i}(u_h)_t(v_h)dx - \int_{I_i}(f(u_h)+r'(u_h)p_h)(v_h)_x dx \\ &+(\widehat{f}+\widehat{r'}\widehat{p})_{i+\frac{1}{2}}(v_h)^-_{i+\frac{1}{2}} - (\widehat{f}+\widehat{r'}\widehat{p})_{i-\frac{1}{2}}(v_h)^+_{i-\frac{1}{2}} \\ &+\int_{I_i} p_h w_h dx \\ &+\int_{I_i} g(q_h)(w_h)_x dx - \widehat{g}_{i+\frac{1}{2}}(w_h)^-_{i+\frac{1}{2}} + \widehat{g}_{i-\frac{1}{2}}(w_h)^+_{i-\frac{1}{2}} \\ &+\int_{I_i} q_h z_h dx \\ &+\int_{I_i} r(u_h)(z_h)_x dx - \widehat{r}_{i+\frac{1}{2}}(z_h)^-_{i+\frac{1}{2}} + \widehat{r}_{i-\frac{1}{2}}(z_h)^+_{i-\frac{1}{2}}. \end{aligned} \tag{5.8}$$

If we take $v_h = u_h$, $w_h = q_h$ and $z_h = -p_h$ in the scheme (5.5), we obtain

$$\begin{aligned} B_i(u_h,p_h,q_h;u_h,q_h,-p_h) = &\int_{I_i}(u_h)_t(u_h)dx \\ &-\int_{I_i}(f(u_h)+r'(u_h)p_h)(u_h)_x dx \\ &+(\widehat{f}+\widehat{r'}\widehat{p})_{i+\frac{1}{2}}(u_h)^-_{i+\frac{1}{2}} \\ &-(\widehat{f}+\widehat{r'}\widehat{p})_{i-\frac{1}{2}}(u_h)^+_{i-\frac{1}{2}} \\ &+\int_{I_i} p_h q_h dx \\ &+\int_{I_i} g(q_h)(q_h)_x dx - \widehat{g}_{i+\frac{1}{2}}(q_h)^-_{i+\frac{1}{2}} + \widehat{g}_{i-\frac{1}{2}}(q_h)^+_{i-\frac{1}{2}} \\ &-\int_{I_i} q_h p_h dx \\ &-\int_{I_i} r(u_h)(p_h)_x dx + \widehat{r}_{i+\frac{1}{2}}(p_h)^-_{i+\frac{1}{2}} - \widehat{r}_{i-\frac{1}{2}}(p_h)^+_{i-\frac{1}{2}} \\ = &\ 0. \end{aligned} \tag{5.9}$$

If we denote $\widetilde{F}(u) = \int^u f(u)du$ and $\widetilde{G}(q) = \int^q g(q)dq$, then (5.9) becomes

$$B_i(u_h,p_h,q_h;u_h,q_h,-p_h) = \frac{1}{2}\frac{d}{dt}\int_{I_i}(u_h)^2\,dx + \widehat{F}_{i+\frac{1}{2}} - \widehat{F}_{i-\frac{1}{2}} + \Theta_{i-\frac{1}{2}} = 0, \tag{5.10}$$

where

$$\widehat{F} = -\widetilde{F}(u_h^-) + \widehat{f}u_h^- + \widetilde{G}(q_h^-) + \widehat{r'}p_h^+ u_h^- - \widehat{g}q_h^-, \tag{5.11}$$

and

$$\Theta = \left(-\widetilde{F}(u_h^-) + \widehat{f}u_h^- + \widetilde{F}(u_h^+) - \widehat{f}u_h^+\right) + \left(\widetilde{G}(q_h^-) - \widehat{g}q_h^- - \widetilde{G}(q_h^+) + \widehat{g}q_h^+\right), \quad (5.12)$$

where we have used the definition of the numerical fluxes (5.6). Notice that we have omitted the subindex $i - \frac{1}{2}$ in the definitions of $\widehat{F}$ and Θ. It is easy to verify that the numerical entropy flux $\widehat{F}$ defined by (5.11) is consistent with the entropy flux $F(u) + ur'(u)p - G(q)$. The terms inside the first parenthesis for Θ in (5.12) are the same as that in (3.16) for the conservation law case; those inside the second parenthesis are the same as those inside the first parenthesis, if we replace q_h by u_h, $-\widetilde{G}$ by $\widetilde{F}$, and $-\widehat{g}$ by $\widehat{f}$ (recall that $-\widehat{g}$ is a monotone flux). We therefore readily have $\Theta \geq 0$. This finishes the proof of (5.7). □

We observe once more that the proof does not depend on the accuracy of the scheme, namely it holds for the piecewise polynomial space (3.8) with any degree k. Also, the same proof can be given for the multi-dimensional LDG scheme solving (5.1) on any triangulation.

As before, the cell entropy inequality trivially implies an L^2-stability of the numerical solution.

Proposition 5.2. *For periodic or compactly supported boundary conditions, the solution u_h to the semi-discrete* LDG *scheme* (5.5) *satisfies the L^2-stability*

$$\frac{d}{dt}\int_0^1 (u_h)^2 dx \leq 0, \quad (5.13)$$

or

$$\|u_h(\cdot, t)\| \leq \|u_h(\cdot, 0)\|. \quad (5.14)$$

□

Again, both the cell entropy inequality (5.7) and the L^2-stability (5.13) are valid regardless of whether the KdV-type equation (5.3) is convection-dominated or dispersion-dominated and regardless of whether the exact solution is smooth or not. The dispersion flux $r'(u)g(r(u)_x)_x$ can be degenerate (equal to zero) in any part of the domain. The LDG method is particularly attractive for convection-dominated convection-dispersion equations, when traditional continuous finite-element methods may be less stable. In [55], this LDG method is used to study the dispersion limit of the Burgers equation, for which the third derivative dispersion term in (5.3) has a small coefficient which tends to zero.

5.1.2 Error Estimates

For error estimates we once again assume the exact solution of (5.3) is smooth. The error estimates can be obtained for a general class of nonlinear convection-dispersion equations which is a subclass of (5.3), see [53]. However, for simplicity we will give here only the proof for the linear equation

$$u_t + u_x + u_{xxx} = 0 \quad (5.15)$$

defined on $[0, 1]$ with periodic boundary conditions.

Proposition 5.3. *The solution u_h to the semi-discrete* LDG *scheme* (5.5) *for the* PDE (5.15) *with a smooth solution u satisfies the following error estimate*

$$\|u - u_h\| \le Ch^{k+\frac{1}{2}}, \tag{5.16}$$

where C depends on u and its derivatives but is independent of h.

Proof. The LDG scheme (5.5), when using the notation in (5.8), can be written as

$$B_i(u_h, p_h, q_h; v_h, w_h, z_h) = 0, \tag{5.17}$$

for all $v_h, w_h, z_h \in V_h$ and for all i. It is easy to verify that the exact solution u, $q = u_x$ and $p = u_{xx}$ of the PDE (5.15) also satisfies

$$B_i(u, p, q; v_h, w_h, z_h) = 0, \tag{5.18}$$

for all $v_h, w_h, z_h \in V_h$ and for all i. Subtracting (5.17) from (5.18) and using the linearity of B_i with respect to its first three arguments, we obtain the error equation

$$B_i(u - u_h, p - p_h, q - q_h; v_h, w_h, z_h) = 0, \tag{5.19}$$

for all $v_h, w_h, z_h \in V_h$ and for all i.

Recall the special projection P defined in (3.37). We also denote the standard L^2-projection as R: for a given smooth function w, the projection Rw is the unique function in V_h which satisfies, for each i,

$$\int_{I_i} (Rw(x) - w(x))v_h(x)dx = 0 \qquad \forall v_h \in P^k(I_i). \tag{5.20}$$

Similar to P, we also have, by the standard approximation theory [7], that

$$\|Rw(x) - w(x)\| + \sqrt{h}\|Rw(x) - w(x)\|_\Gamma \le Ch^{k+1} \tag{5.21}$$

for a smooth function w, where C is a constant depending on w and its derivatives but independent of h, and $\|v\|_\Gamma$ is the usual L^2-norm on the cell interfaces of the mesh, which for this one-dimensional case is

$$\|v\|_\Gamma^2 = \sum_i \left((v^-_{i+\frac{1}{2}})^2 + (v^+_{i-\frac{1}{2}})^2 \right).$$

We now take

$$v_h = Pu - u_h, \qquad w_h = Rq - q_h, \qquad z_h = p_h - Rp \tag{5.22}$$

in the error equation (5.19), and denote

$$\begin{aligned} &e_h = Pu - u_h, \;\; \bar{e}_h = Rq - q_h, \\ &\bar{\bar{e}}_h = Rp - p_h; \quad \varepsilon_h = u - Pu, \;\; \bar{\varepsilon}_h = q - Rq, \;\; \bar{\bar{\varepsilon}}_h = p - Rp, \end{aligned} \tag{5.23}$$

to obtain

$$B_i(e_h,\bar{\bar{e}}_h,\bar{e}_h,;e_h,\bar{e}_h,-\bar{\bar{e}}_h) = -B_i(\varepsilon_h,\bar{\bar{\varepsilon}}_h,\bar{\varepsilon}_h;e_h,\bar{e}_h,-\bar{\bar{e}}_h). \tag{5.24}$$

For the left-hand side of (5.24), we use the cell entropy inequality (see (5.10)) to obtain

$$B_i(e_h,\bar{\bar{e}}_h,\bar{e}_h,;e_h,\bar{e}_h,-\bar{\bar{e}}_h) = \frac{1}{2}\frac{d}{dt}\int_{I_i}(e_h)^2dx + \widehat{F}_{i+\frac{1}{2}} - \widehat{F}_{i-\frac{1}{2}} + \Theta_{i-\frac{1}{2}} \tag{5.25}$$

where we can easily verify, based on the formula (5.12) and for the PDE (5.15), that

$$\Theta_{i-\frac{1}{2}} = \frac{1}{2}\left((e_h)^+_{i-\frac{1}{2}} - (e_h)^-_{i-\frac{1}{2}}\right)^2 + \frac{1}{2}\left((\bar{e}_h)^+_{i-\frac{1}{2}} - (\bar{e}_h)^-_{i-\frac{1}{2}}\right)^2. \tag{5.26}$$

As to the right-hand side of (5.24), we first write out all the terms

$$\begin{aligned}
-B_i(&\varepsilon_h,\bar{\bar{\varepsilon}}_h,\bar{\varepsilon}_h;e_h,\bar{e}_h,-\bar{\bar{e}}_h)\\
&= -\int_{I_i}(\varepsilon_h)_t e_h dx\\
&+ \int_{I_i}(\varepsilon_h+\bar{\bar{\varepsilon}}_h)(e_h)_x dx - (\varepsilon_h^- + \bar{\bar{\varepsilon}}_h^+)_{i+\frac{1}{2}}(e_h)^-_{i+\frac{1}{2}} + (\varepsilon_h^- + \bar{\bar{\varepsilon}}_h^+)_{i-\frac{1}{2}}(e_h)^+_{i-\frac{1}{2}}\\
&- \int_{I_i}\bar{\bar{\varepsilon}}_h\bar{e}_h dx - \int_{I_i}\bar{\varepsilon}_h(\bar{e}_h)_x dx + (\bar{\varepsilon}_h)^+_{i+\frac{1}{2}}(\bar{e}_h)^-_{i+\frac{1}{2}} - (\bar{\varepsilon}_h)^+_{i-\frac{1}{2}}(\bar{e}_h)^+_{i-\frac{1}{2}}\\
&+ \int_{I_i}\bar{\varepsilon}_h\bar{\bar{e}}_h dx + \int_{I_i}\varepsilon_h(\bar{\bar{e}}_h)_x dx - (\varepsilon_h)^-_{i+\frac{1}{2}}(\bar{\bar{e}}_h)^-_{i+\frac{1}{2}} + (\varepsilon_h)^-_{i-\frac{1}{2}}(\bar{\bar{e}}_h)^+_{i-\frac{1}{2}}.
\end{aligned}$$

Noticing the properties (3.37) and (5.20) of the projections P and R, we have

$$\int_{I_i}(\varepsilon_h+\bar{\bar{\varepsilon}}_h)(e_h)_x dx = 0, \qquad \int_{I_i}\bar{\bar{\varepsilon}}_h\bar{e}_h dx = 0, \qquad \int_{I_i}\bar{\varepsilon}_h(\bar{e}_h)_x dx = 0,$$

$$\int_{I_i}\bar{\varepsilon}_h\bar{\bar{e}}_h dx = 0, \qquad \int_{I_i}\varepsilon_h(\bar{\bar{e}}_h)_x dx = 0,$$

because $(e_h)_x$, $(\bar{e}_h)_x$ and $(\bar{\bar{e}}_h)_x$ are polynomials of degree at most $k-1$, and $\bar{e}_h$ and $\bar{\bar{e}}_h$ are polynomials of degree at most k. Also,

$$(\varepsilon_h)^-_{i+\frac{1}{2}} = u_{i+\frac{1}{2}} - (Pu)^-_{i+\frac{1}{2}} = 0$$

for all i. Therefore, the right-hand side of (5.24) becomes

$$\begin{aligned}
-B_i(&\varepsilon_h,\bar{\bar{\varepsilon}}_h,\bar{\varepsilon}_h;e_h,\bar{e}_h,-\bar{\bar{e}}_h)\\
&= -\int_{I_i}(\varepsilon_h)_t e_h dx - (\bar{\bar{\varepsilon}}_h)^+_{i+\frac{1}{2}}(e_h)^-_{i+\frac{1}{2}} + (\bar{\bar{\varepsilon}}_h)^+_{i-\frac{1}{2}}(e_h)^+_{i-\frac{1}{2}}\\
&\qquad + (\bar{\varepsilon}_h)^+_{i+\frac{1}{2}}(\bar{e}_h)^-_{i+\frac{1}{2}} - (\bar{\varepsilon}_h)^+_{i-\frac{1}{2}}(\bar{e}_h)^+_{i-\frac{1}{2}}
\end{aligned}$$

$$
\begin{aligned}
&= -\int_{I_i} (\varepsilon_h)_t e_h dx + \widehat{H}_{i+\frac{1}{2}} - \widehat{H}_{i-\frac{1}{2}} \\
&\quad + (\bar{\bar{\varepsilon}}_h)^+_{i-\frac{1}{2}} \left((e_h)^+_{i-\frac{1}{2}} - (e_h)^-_{i-\frac{1}{2}} \right) - (\bar{\varepsilon}_h)^+_{i-\frac{1}{2}} \left((\bar{e}_h)^+_{i-\frac{1}{2}} - (\bar{e}_h)^-_{i-\frac{1}{2}} \right) \qquad (5.27) \\
&\le \widehat{H}_{i+\frac{1}{2}} - \widehat{H}_{i-\frac{1}{2}} + \frac{1}{2} \Bigg[\int_{I_i} ((\varepsilon_h)_t)^2 dx + \int_{I_i} (e_h)^2 dx \\
&\qquad + \left((\bar{\bar{\varepsilon}}_h)^+_{i-\frac{1}{2}} \right)^2 + \left((e_h)^+_{i-\frac{1}{2}} - (e_h)^-_{i-\frac{1}{2}} \right)^2 \\
&\qquad + \left((\bar{\varepsilon}_h)^+_{i-\frac{1}{2}} \right)^2 + \left((\bar{e}_h)^+_{i-\frac{1}{2}} - (\bar{e}_h)^-_{i-\frac{1}{2}} \right)^2 \Bigg].
\end{aligned}
$$

Plugging (5.25), (5.26) and (5.27) into the equality (5.24), summing up over i, and using the approximation results (3.38) and (5.21), we obtain

$$
\frac{d}{dt} \int_0^1 (e_h)^2 dx \le \int_0^1 (e_h)^2 dx + Ch^{2k+1}.
$$

A Gronwall's inequality, the fact that the initial error

$$
\| u(\cdot, 0) - u_h(\cdot, 0) \| \le Ch^{k+1},
$$

and the approximation results (3.38) and (5.21) finally give us the error estimate (5.16). □

We note that the error estimate (5.16) is half an order lower than optimal. Technically, this is because we are unable to use the special projections as before to eliminate the interface terms involving $\bar{\varepsilon}_h$ and $\bar{\bar{\varepsilon}}_h$ in (5.27). Numerical experiments in [55] indicate that both the L^2- and L^∞-errors are of the optimal $(k+1)$-th order of accuracy.

5.2 LDG Schemes for Other Higher-Order PDEs

In this subsection we list some of the higher-order PDEs for which stable DG methods have been designed in the literature. We will concentrate on the discussion of LDG schemes.

5.2.1 Bi-harmonic Equations

An LDG scheme for solving the time-dependent convection-bi-harmonic equation

$$
u_t + \sum_{i=1}^{d} f_i(u)_{x_i} + \sum_{i=1}^{d} (a_i(u_{x_i}) u_{x_i x_i})_{x_i x_i} = 0, \qquad (5.28)
$$

where $f_i(u)$ and $a_i(q) \geq 0$ are arbitrary functions, was designed in [56]. The numerical fluxes are chosen following the same "alternating fluxes" principle similar to the second-order convection-diffusion equation (4.1), see (4.6). A cell entropy inequality and the L^2-stability of the LDG scheme for the nonlinear equation (5.28) can be proved [56], which do not depend on the smoothness of the solution of (5.28), the order of accuracy of the scheme, or the triangulation.

5.2.2 Fifth-Order Convection-Dispersion Equations

An LDG scheme for solving the following fifth-order convection-dispersion equation

$$u_t + \sum_{i=1}^{d} f_i(u)_{x_i} + \sum_{i=1}^{d} g_i(u_{x_i x_i})_{x_i x_i x_i} = 0, \tag{5.29}$$

where $f_i(u)$ and $g_i(q)$ are arbitrary functions, was designed in [56]. The numerical fluxes are chosen following the same upwinding and "alternating fluxes" principle similar to the third-order KdV-type equations (5.1), see (5.6). A cell entropy inequality and the L^2-stability of the LDG scheme for the nonlinear equation (5.29) can be proved [56], which again do not depend on the smoothness of the solution of (5.29), the order of accuracy of the scheme, or the triangulation.

Stable LDG schemes for similar equations with sixth or higher derivatives can also be designed along similar lines.

5.2.3 The $K(m, n)$ Equations

LDG methods for solving the $K(m, n)$ equations

$$u_t + (u^m)_x + (u^n)_{xxx} = 0, \tag{5.30}$$

where m and n are positive integers, have been designed in [27]. These $K(m, n)$ equations were introduced by Rosenau and Hyman in [40] to study the so-called *compactons*, namely the compactly supported solitary waves solutions. For the special case of $m = n$ being an odd positive integer, LDG schemes which are stable in the L^{m+1}-norm can be designed (see [27]). For other cases, we can also design LDG schemes based on a linearized stability analysis, which perform well in numerical simulation for the fully nonlinear equation (5.30).

5.2.4 The KdV-Burgers-Type (KdVB) Equations

LDG methods for solving the KdV-Burgers-type (KdVB) equations

$$u_t + f(u)_x - (a(u)u_x)_x + (r'(u)g(r(u)_x)_x)_x = 0, \tag{5.31}$$

where $f(u)$, $a(u) \geq 0$, $r(u)$ and $g(q)$ are arbitrary functions, have been designed in [49]. The design of numerical fluxes follows the same lines as that for the convection-diffusion equation (4.2) and for the KdV-type equation (5.3). A cell entropy inequality and the L^2-stability of the LDG scheme for the nonlinear equation (5.31) can be proved [49], which again do not depend on the smoothness of the solution of (5.31) and the order of accuracy of the scheme. The LDG scheme is used in [49] to study different regimes when one of the dissipation and the dispersion mechanisms dominates, and when they have comparable influence on the solution. An advantage of the LDG scheme designed in [49] is that it is stable regardless of which mechanism (convection, diffusion, dispersion) actually dominates.

5.2.5 The Fifth-Order KdV-Type Equations

LDG methods for solving the fifth-order KdV-type equations

$$u_t + f(u)_x + (r'(u)g(r(u)_x)_x)_x + (s'(u)h(s(u)_{xx})_{xx})_x = 0, \tag{5.32}$$

where $f(u)$, $r(u)$, $g(q)$, $s(u)$ and $h(p)$ are arbitrary functions, have been designed in [49]. The design of numerical fluxes follows the same lines as that for the KdV-type equation (5.3). A cell entropy inequality and the L^2-stability of the LDG scheme for the nonlinear equation (5.32) can be proved [49], which again do not depend on the smoothness of the solution of (5.32) and the order of accuracy of the scheme. The LDG scheme is used in [49] to simulate the solutions of the Kawahara equation, the generalized Kawahara equation, Ito's fifth-order KdV equation, and a fifth-order KdV-type equations with high nonlinearities, which are all special cases of the equations represented by (5.32).

5.2.6 The Fully Nonlinear $K(n, n, n)$ Equations

LDG methods for solving the fifth-order fully nonlinear $K(n, n, n)$ equations

$$u_t + (u^n)_x + (u^n)_{xxx} + (u^n)_{xxxxx} = 0, \tag{5.33}$$

where n is a positive integer, have been designed in [49]. The design of numerical fluxes follows the same lines as that for the $K(m, n)$ equations (5.30). For odd n, stability in the L^{n+1}-norm of the resulting LDG scheme can be proved for the nonlinear equation (5.33) [49]. This scheme is used to simulate compacton propagation in [49].

5.2.7 The Nonlinear Schrödinger (NLS) Equation

In [50], LDG methods are designed for the generalized nonlinear Schrödinger (NLS) equation

$$i\,u_t + u_{xx} + i\,(g(|u|^2)u)_x + f(|u|^2)u = 0, \tag{5.34}$$

the two-dimensional version

$$i\,u_t + \Delta u + f(|u|^2)u = 0, \tag{5.35}$$

and the coupled nonlinear Schrödinger equation

$$\begin{cases} i\,u_t + i\,\alpha u_x + u_{xx} + \beta\,u + \kappa v + f(|u|^2, |v|^2)u = 0 \\ i\,v_t - i\,\alpha v_x + v_{xx} - \beta\,u + \kappa v + g(|u|^2, |v|^2)v = 0, \end{cases} \tag{5.36}$$

where $f(q)$ and $g(q)$ are arbitrary functions and α, β and κ are constants. With suitable choices of the numerical fluxes, the resulting LDG schemes are proved to satisfy a cell entropy inequality and L^2-stability [50]. The LDG scheme is used in [50] to simulate the soliton propagation and interaction, and the appearance of singularities. The easiness of h-p adaptivity of the LDG scheme and rigorous stability for the fully nonlinear case make it an ideal choice for the simulation of Schrödinger equations, for which the solutions often have quite localized structures.

5.2.8 The Kadomtsev-Petviashvili (KP) Equations

The two-dimensional Kadomtsev-Petviashvili (KP) equations

$$(u_t + 6uu_x + u_{xxx})_x + 3\sigma^2 u_{yy} = 0, \tag{5.37}$$

where $\sigma^2 = \pm 1$, are generalizations of the one-dimensional KdV equations and are important models for water waves. Because of the x-derivative for the u_t term, the equation (5.37) is well posed only in a function space with a global constraint, hence it is very difficult to design an efficient LDG scheme which relies on local operations. In [51], an LDG scheme for (5.37) is designed by carefully choosing locally supported bases which satisfy the global constraint needed by the solution of (5.37). The LDG scheme satisfies a cell entropy inequality and is L^2-stable for the fully nonlinear equation (5.37). Numerical simulations are performed in [51] for both the KP-I equations ($\sigma^2 = -1$ in (5.37)) and the KP-II equations ($\sigma^2 = 1$ in (5.37)). Line solitons and lump-type pulse solutions have been simulated.

5.2.9 The Zakharov-Kuznetsov (ZK) Equation

The two-dimensional Zakharov-Kuznetsov (ZK) equation

$$u_t + (3u^2)_x + u_{xxx} + u_{xyy} = 0 \tag{5.38}$$

is another generalization of the one-dimensional KdV equations. An LDG scheme is designed for (5.38) in [51] which is proved to satisfy a cell entropy inequality and to be L^2-stable. An L^2-error estimate is given in [53]. Various nonlinear waves have been simulated by this scheme in [51].

5.2.10 The Kuramoto-Sivashinsky-type Equations

In [52], an LDG method is developed to solve the Kuramoto-Sivashinsky-type equations

$$u_t + f(u)_x - (a(u)u_x)_x + (r'(u)g(r(u)_x)_x)_x + (s(u_x)u_{xx})_{xx} = 0, \tag{5.39}$$

where $f(u)$, $a(u)$, $r(u)$, $g(q)$ and $s(p) \geq 0$ are arbitrary functions. The Kuramoto-Sivashinsky equation

$$u_t + uu_x + \alpha u_{xx} + \beta u_{xxxx} = 0, \tag{5.40}$$

where α and $\beta \geq 0$ are constants, which is a special case of (5.39), is a canonical evolution equation which has attracted considerable attention over the last decades. When the coefficients α and β are both positive, its linear terms describe a balance between long-wave instability and short-wave stability, with the nonlinear term providing a mechanism for energy transfer between wave modes. The LDG method developed in [52] can be proved to satisfy a cell entropy inequality and is therefore L^2-stable, for the general nonlinear equation (5.39). The LDG scheme is used in [52] to simulate chaotic solutions of (5.40).

5.2.11 The Ito-Type Coupled KdV Equations

Also in [52], an LDG method is developed to solve the Ito-type coupled KdV equations

$$\begin{aligned} u_t + \alpha uu_x + \beta vv_x + \gamma u_{xxx} &= 0, \\ v_t + \beta(uv)_x &= 0, \end{aligned} \tag{5.41}$$

where α, β and γ are constants. An L^2-stability is proved for the LDG method. Simulation for the solution of (5.41) in which the result for u behaves like dispersive wave solution and the result for v behaves like shock wave solution is performed in [52] using the LDG scheme.

5.2.12 The Camassa-Holm (CH) Equation

An LDG method for solving the Camassa-Holm (CH) equation

$$u_t - u_{xxt} + 2\kappa u_x + 3uu_x = 2u_x u_{xx} + uu_{xxx}, \tag{5.42}$$

where κ is a constant, is designed in [54]. Because of the u_{xxt} term, the design of an LDG method is non-standard. By a careful choice of the numerical fluxes, the authors obtain an LDG scheme which can be proved to satisfy a cell entropy inequality and to be L^2-stable [54]. A sub-optimal $O(h^k)$ error estimate is also obtained in [54].

5.2.13 The Cahn-Hilliard Equation

LDG methods have been designed for solving the Cahn-Hilliard equation

$$u_t = \nabla \cdot \Big(b(u) \nabla \big(-\gamma \Delta u + \Psi'(u) \big) \Big), \tag{5.43}$$

and the Cahn-Hilliard system

$$\begin{cases} \boldsymbol{u}_t &= \nabla \cdot (\boldsymbol{B}(\boldsymbol{u})\nabla \boldsymbol{\omega}), \\ \boldsymbol{\omega} &= -\gamma \Delta \boldsymbol{u} + D\Psi(\boldsymbol{u}), \end{cases} \tag{5.44}$$

in [47], where $\{D\Psi(\boldsymbol{u})\}_l = \frac{\partial \Psi(\boldsymbol{u})}{\partial u_l}$ and γ is a positive constant. Here $b(u)$ is the non-negative diffusion mobility and $\Psi(u)$ is the homogeneous free energy density for the scalar case (5.43). For the system case (5.44), $\boldsymbol{B}(\boldsymbol{u})$ is the symmetric positive semi-definite mobility matrix and $\Psi(\boldsymbol{u})$ is the homogeneous free energy density. The proof of the energy stability for the LDG scheme is given for the general nonlinear solutions. Many simulation results are given in [47].

In [48], a class of LDG methods are designed for the more general Allen-Cahn/Cahn-Hilliard (AC/CH) system in $\Omega \in \mathbb{R}^d$ $(d \leq 3)$

$$\begin{cases} u_t &= \nabla \cdot [b(u,v)\nabla(\Psi_u(u,v) - \gamma \Delta u)], \\ \rho v_t &= -b(u,v)[\Psi_v(u,v) - \gamma \Delta v]. \end{cases} \tag{5.45}$$

Energy stability of the LDG schemes is again proved. Simulation results are provided.

Bibliography

[1] D. Arnold, F. Brezzi, B. Cockburn and L. Marini, Unified analysis of discontinuous Galerkin methods for elliptic problems. *SIAM Journal on Numerical Analysis* **39** (2002), 1749–1779.

[2] H. Atkins and C.-W. Shu, Quadrature-free implementation of the discontinuous Galerkin method for hyperbolic equations. *AIAA Journal* **36** (1998), 775–782.

[3] F. Bassi and S. Rebay, A high-order accurate discontinuous finite element method for the numerical solution of the compressible Navier-Stokes equations. *Journal of Computational Physics* **131** (1997), 267–279.

[4] C.E. Baumann and J.T. Oden, A discontinuous *hp* finite element method for convection-diffusion problems. *Computer Methods in Applied Mechanics and Engineering* **175** (1999), 311–341.

[5] R. Biswas, K.D. Devine and J. Flaherty, Parallel, adaptive finite element methods for conservation laws. *Applied Numerical Mathematics* **14** (1994), 255–283.

[6] Y. Cheng and C.-W. Shu, A discontinuous Galerkin finite element method for time-dependent partial differential equations with higher order derivatives. *Mathematics of Computation* **77** (2008), 699–730.

[7] P. Ciarlet, *The Finite Element Method for Elliptic Problems.* North Holland, 1975.

[8] B. Cockburn, Discontinuous Galerkin methods for convection-dominated problems. In: *High-Order Methods for Computational Physics*, T.J. Barth and H. Deconinck, editors, Lecture Notes in Computational Science and Engineering, volume 9, Springer, 1999, 69–224.

[9] B. Cockburn, B. Dong and J. Guzmán, Optimal convergence of the original DG method for the transport-reaction equation on special meshes. *SIAM Journal on Numerical Analysis* **46** (2008), 1250–1265.

[10] B. Cockburn, S. Hou and C.-W. Shu, The Runge-Kutta local projection discontinuous Galerkin finite element method for conservation laws IV: the multidimensional case. *Mathematics of Computation* **54** (1990), 545–581.

[11] B. Cockburn, G. Karniadakis and C.-W. Shu, The development of discontinuous Galerkin methods. In: *Discontinuous Galerkin Methods: Theory, Computation and Applications*, B. Cockburn, G. Karniadakis and C.-W. Shu, editors, Lecture Notes in Computational Science and Engineering, volume 11, Springer, 2000, Part I: Overview, 3–50.

[12] B. Cockburn, S.-Y. Lin and C.-W. Shu, TVB Runge-Kutta local projection discontinuous Galerkin finite element method for conservation laws III: one dimensional systems. *Journal of Computational Physics* **84** (1989), 90–113.

[13] B. Cockburn and C.-W. Shu, TVB Runge-Kutta local projection discontinuous Galerkin finite element method for conservation laws II: general framework. *Mathematics of Computation* **52** (1989), 411–435.

[14] B. Cockburn and C.-W. Shu, The Runge-Kutta local projection P^1-discontinuous-Galerkin finite element method for scalar conservation laws. *Mathematical Modelling and Numerical Analysis* (M^2AN) **25** (1991), 337–361.

[15] B. Cockburn and C.-W. Shu, The Runge-Kutta discontinuous Galerkin method for conservation laws V: multidimensional systems. *Journal of Computational Physics* **141** (1998), 199–224.

[16] B. Cockburn and C.-W. Shu, The local discontinuous Galerkin method for time-dependent convection-diffusion systems. *SIAM Journal on Numerical Analysis* **35** (1998), 2440–2463.

[17] B. Cockburn and C.-W. Shu, Runge-Kutta Discontinuous Galerkin methods for convection-dominated problems. *Journal of Scientific Computing* **16** (2001), 173–261.

[18] B. Cockburn and C.-W. Shu, Foreword for the special issue on discontinuous Galerkin method. *Journal of Scientific Computing* **22–23** (2005), 1–3.

[19] C. Dawson, Foreword for the special issue on discontinuous Galerkin method. *Computer Methods in Applied Mechanics and Engineering* **195** (2006), 3183.

[20] S. Gottlieb and C.-W. Shu, Total variation diminishing Runge-Kutta schemes. *Mathematics of Computation* **67** (1998), 73–85.

[21] S. Gottlieb, C.-W. Shu and E. Tadmor, Strong stability preserving high order time discretization methods. *SIAM Reviews* **43** (2001), 89–112.

[22] A. Harten, High resolution schemes for hyperbolic conservation laws. *Journal of Computational Physics* **49** (1983), 357–393.

[23] G.-S. Jiang and C.-W. Shu, On cell entropy inequality for discontinuous Galerkin methods. *Mathematics of Computation* **62** (1994), 531–538.

[24] C. Johnson and J. Pitkäranta, An analysis of the discontinuous Galerkin method for a scalar hyperbolic equation. *Mathematics of Computation* **46** (1986), 1–26.

[25] P. Lesaint and P.A. Raviart, On a finite element method for solving the neutron transport equation. In: *Mathematical aspects of finite elements in partial differential equations*, C. de Boor, ed., Academic Press, 1974, 89–145.

[26] R.J. LeVeque, *Numerical Methods for Conservation Laws*. Birkhäuser, Basel, 1990.

[27] D. Levy, C.-W. Shu and J. Yan, Local discontinuous Galerkin methods for nonlinear dispersive equations. *Journal of Computational Physics* **196** (2004), 751–772.

[28] P.-L. Lions and P.E. Souganidis, Convergence of MUSCL and filtered schemes for scalar conservation law and Hamilton-Jacobi equations. *Numerische Mathematik* **69** (1995), 441–470.

[29] J.T. Oden, I. Babuvska and C.E. Baumann, A discontinuous *hp* finite element method for diffusion problems. *Journal of Computational Physics* **146** (1998), 491–519.

[30] S. Osher, Convergence of generalized MUSCL schemes. *SIAM Journal on Numerical Analysis* **22** (1985), 947–961.

[31] S. Osher and S. Chakravarthy, High resolution schemes and the entropy condition. *SIAM Journal on Numerical Analysis* **21** (1984), 955–984.

[32] S. Osher and E. Tadmor, On the convergence of the difference approximations to scalar conservation laws. *Mathematics of Computation* **50** (1988), 19–51.

[33] T. Peterson, A note on the convergence of the discontinuous Galerkin method for a scalar hyperbolic equation. *SIAM Journal on Numerical Analysis* **28** (1991), 133–140.

[34] J. Qiu and C.-W. Shu, Hermite WENO schemes and their application as limiters for Runge-Kutta discontinuous Galerkin method: one dimensional case. *Journal of Computational Physics* **193** (2003), 115–135.

[35] J. Qiu and C.-W. Shu, Runge-Kutta discontinuous Galerkin method using WENO limiters. *SIAM Journal on Scientific Computing* **26** (2005), 907–929.

[36] J. Qiu and C.-W. Shu, Hermite WENO schemes and their application as limiters for Runge-Kutta discontinuous Galerkin method II: two dimensional case. *Computers & Fluids* **34** (2005), 642–663.

[37] W.H. Reed and T.R. Hill, Triangular mesh methods for the neutron transport equation. Tech. Report LA-UR-73-479, Los Alamos Scientific Laboratory, 1973.

[38] J.-F. Remacle, J. Flaherty and M. Shephard, An adaptive discontinuous Galerkin technique with an orthogonal basis applied to Rayleigh-Taylor flow instabilities. *SIAM Review* **45** (2003), 53–72.

[39] G.R. Richter, An optimal-order error estimate for the discontinuous Galerkin method. *Mathematics of Computation* **50** (1988), 75–88.

[40] P. Rosenau and J.M. Hyman, Compactons: solitons with finite wavelength. *Physical Review Letters* **70** (1993), 564–567.

[41] C.-W. Shu, TVB uniformly high-order schemes for conservation laws. *Mathematics of Computation* **49** (1987), 105–121.

[42] C.-W. Shu, Total-Variation-Diminishing time discretizations. *SIAM Journal on Scientific and Statistical Computing* **9** (1988), 1073–1084.

[43] C.-W. Shu, A survey of strong stability preserving high order time discretizations. In: *Collected Lectures on the Preservation of Stability under Discretization*, D. Estep and S. Tavener, editors, SIAM, 2002, 51–65.

[44] C.-W. Shu and S. Osher, Efficient implementation of essentially non-oscillatory shock-capturing schemes. *Journal of Computational Physics* **77** (1988), 439–471.

[45] B. van Leer and S. Nomura, Discontinuous Galerkin for diffusion. 17th AIAA Computational Fluid Dynamics Conference (June 6–9, 2005), AIAA paper 2005–5108.

[46] Y. Xia, Y. Xu and C.-W. Shu, Efficient time discretization for local discontinuous Galerkin methods. *Discrete and Continuous Dynamical Systems – Series B* **8** (2007), 677–693.

[47] Y. Xia, Y. Xu and C.-W. Shu, Local discontinuous Galerkin methods for the Cahn-Hilliard type equations. *Journal of Computational Physics* **227** (2007), 472–491.

[48] Y. Xia, Y. Xu and C.-W. Shu, Application of the local discontinuous Galerkin method for the Allen-Cahn/Cahn-Hilliard system. *Communications in Computational Physics* **5** (2009), 821–835.

[49] Y. Xu and C.-W. Shu, Local discontinuous Galerkin methods for three classes of nonlinear wave equations. *Journal of Computational Mathematics* **22** (2004), 250–274.

[50] Y. Xu and C.-W. Shu, Local discontinuous Galerkin methods for nonlinear Schrödinger equations. *Journal of Computational Physics* **205** (2005), 72–97.

[51] Y. Xu and C.-W. Shu, Local discontinuous Galerkin methods for two classes of two dimensional nonlinear wave equations. *Physica D* **208** (2005), 21–58.

[52] Y. Xu and C.-W. Shu, Local discontinuous Galerkin methods for the Kuramoto-Sivashinsky equations and the Ito-type coupled KdV equations. *Computer Methods in Applied Mechanics and Engineering* **195** (2006), 3430–3447.

[53] Y. Xu and C.-W. Shu, Error estimates of the semi-discrete local discontinuous Galerkin method for nonlinear convection-diffusion and KdV equations. *Computer Methods in Applied Mechanics and Engineering* **196** (2007), 3805–3822.

[54] Y. Xu and C.-W. Shu, A local discontinuous Galerkin method for the Camassa-Holm equation. *SIAM Journal on Numerical Analysis* **46** (2008), 1998–2021.

[55] J. Yan and C.-W. Shu, A local discontinuous Galerkin method for KdV type equations. *SIAM Journal on Numerical Analysis* **40** (2002), 769–791.

[56] J. Yan and C.-W. Shu, Local discontinuous Galerkin methods for partial differential equations with higher order derivatives. *Journal of Scientific Computing* **17** (2002), 27–47.

[57] M. Zhang and C.-W. Shu, An analysis of three different formulations of the discontinuous Galerkin method for diffusion equations. *Mathematical Models and Methods in Applied Sciences* (M^3AS) **13** (2003), 395–413.

[58] Q. Zhang and C.-W. Shu, Error estimates to smooth solutions of Runge-Kutta discontinuous Galerkin methods for scalar conservation laws. *SIAM Journal on Numerical Analysis* **42** (2004), 641–666.

[59] Q. Zhang and C.-W. Shu, Error estimates to smooth solutions of Runge-Kutta discontinuous Galerkin method for symmetrizable systems of conservation laws. *SIAM Journal on Numerical Analysis* **44** (2006), 1703–1720.

[60] J. Zhu, J.-X. Qiu, C.-W. Shu and M. Dumbser, Runge-Kutta discontinuous Galerkin method using WENO limiters II: unstructured meshes. *Journal of Computational Physics* **227** (2008), 4330–4353.

Advanced Courses in Mathematics CRM Barcelona

Edited by
Manuel Castellet

BIRKHÄUSER

Since 1995 the Centre de Recerca Matemàtica (CRM) in Barcelona has conducted a number of annual Summer Schools at the post-doctoral or advanced graduate level. Sponsored mainly by the European Community, these Advanced Courses have usually been held at the CRM in Bellaterra.
The books in this series consist essentially of the expanded and embellished material presented by the authors in their lectures.

Argyros, S. / Todorcevic, S.
Ramsey Methods in Analysis (2005)
ISBN 978-3-7643-7264-4

Audin, M. / Cannas da Silva, A. / Lerman, E.
Symplectic Geometry of Integrable Hamiltonian Systems (2003)
ISBN 978-3-7643-2167-3

Bertoluzza, S. / Falletta, S. / Russo, G. / Shu, C.-W.
Numerical Solutions of Partial Differential Equations (2008)
ISBN 978-3-7643-8939-0

Brady, N. / Riley, T. / Short, H.
The Geometry of the Word Problem for Finitely Generated Groups (2006)
ISBN 978-3-7643-7949-0

The origins of the word problem are in group theory, decidability and complexity, but, through the vision of Gromov and the language of filling functions, the topic now impacts the world of large-scale geometry, including topics such as soap films, isoperimetry, coarse invariants and curvature.
The first part introduces van Kampen diagrams in Cayley graphs of finitely generated, infinite groups; it discusses the van Kampen lemma, the isoperimetric functions or Dehn functions, the theory of small cancellation groups and an introduction to hyperbolic groups. The second part is dedicated to Dehn functions, negatively curved groups, in particular, CAT(0) groups, cubings and cubical complexes. In the last part, filling functions are presented from geometric, algebraic and algorithmic points of view. Many examples and open problems are included.

Brown, K.A. / Goodearl, K.R.
Lectures on Algebraic Quantum Groups (2002)
ISBN 978-3-7643-6714-5

Catalano, D. / Cramer, R. / Damgård, I. / Di Creszenso, G. / Pointcheval, D. / Takagi, T.
Contemporary Cryptology (2005)
ISBN 978-3-7643-7294-1

Christopher, C. / Li, C.
Limit Cycles of Differential Equations (2007)
ISBN 978-3-7643-8409-8

Cohen, R.L. / Hess, K. / Voronov, A.A.
String Topology and Cyclic Homology (2006)
ISBN 978-3-7643-2182-6

Da Prato, G.
Kolmogorov Equations for Stochastic PDEs (2004)
ISBN 978-3-7643-7216-3

Drensky, V. / Formanek, E.
Polynomial Identity Rings (2004)
ISBN 978-3-7643-7126-5

Dwyer, W.G. / Henn, H.-W.
Homotopy Theoretic Methods in Group Cohomology (2001)
ISBN 978-3-7643-6605-6

Markvorsen, S. / Min-Oo, M.
Global Riemannian Geometry: Curvature and Topology (2003)
ISBN 978-3-7643-2170-3

Mislin, G. / Valette, A.
Proper Group Actions and the Baum-Connes Conjecture (2003)
ISBN 978-3-7643-0408-9

Myasnikov, A. / Shpilrain, V. / Ushakov, A.
Group-based Cryptography (2008)
ISBN 978-3-7643-8826-3

This book is about relations between three different areas of mathematics and theoretical computer science: combinatorial group theory, cryptography, and complexity theory. It is explored how non-commutative (infinite) groups, which are typically studied in combinatorial group theory, can be used in public key cryptography. It is also shown that there is a remarkable feedback from cryptography to combinatorial group theory because some of the problems motivated by cryptography appear to be new to group theory, and they open many interesting research avenues within group theory.
Then, complexity theory, notably generic-case complexity of algorithms, is employed for cryptanalysis of various cryptographic protocols based on infinite groups, and the ideas and machinery from the theory of generic-case complexity are used to study asymptotically dominant properties of some infinite groups that have been applied in public key cryptography so far.
Its elementary exposition makes the book accessible to graduate as well as undergraduate students in mathematics or computer science.